T0202865

Universitext

Thomas E. Cecil

Lie Sphere Geometry

With Applications to Submanifolds

Second Edition

 Springer

Thomas E. Cecil
Department of Mathematics and Computer Science
College of the Holy Cross
1 College Street
Worcester, MA 01610
cecil@mathcs.holycross.edu

ISBN: 978-0-387-74655-5   e-ISBN: 978-0-387-74656-2

Library of Congress Control Number: 2007936690

Mathematics Subject Classification (2000): 53-02, 53A07, 53A40, 53B25, 53C40, 53C42

To my sons,
Tom, Mark, and Michael

Preface to the First Edition

The purpose of this monograph is to provide an introduction to Lie's geometry of oriented spheres and its recent applications to the study of submanifolds of Euclidean space. Lie [104] introduced his sphere geometry in his dissertation, published as a paper in 1872, and used it in his study of contact transformations. The subject was actively pursued through the early part of the twentieth century, culminating with the publication in 1929 of the third volume of Blaschke's [10] *Vorlesungen über Differentialgeometrie*, which is devoted entirely to Lie sphere geometry and its subgeometries. After this, the subject fell out of favor until 1981, when Pinkall [146] used it as the principal tool in his classification of Dupin hypersurfaces in $\mathbf{R}^4$. Since that time, it has been employed by several geometers in the study of Dupin, isoparametric and taut submanifolds.

This book is not intended to replace Blaschke's work, which contains a wealth of material, particularly in dimensions two and three. Rather, it is meant to be a relatively brief introduction to the subject, which leads the reader to the frontiers of current research in this part of submanifold theory. Chapters 2 and 3 (chapter numbers from the second edition) are accessible to a beginning graduate student who has taken courses in linear and abstract algebra and projective geometry. Chapters 4 and 5 contain the applications to submanifold theory. These chapters require a first graduate course in differential geometry as a necessary background. A detailed description of the contents of the individual chapters is given in the introduction, which also serves as a survey of the field to this point in time.

I wish to acknowledge certain works which have been especially useful to me in writing this book. Much of Chapters 2 and 3 is based on Blaschke's book. The proof of the Cartan and Dieudonné theorem in Section 3.2 is taken from E. Artin's book [4], *Geometric Algebra*. Two sources are particularly influential in Chapters 4 and 5. The first is Pinkall's dissertation [146] and his subsequent paper [150], which have proven to be remarkably fruitful. Secondly, the approach to the study of Legendre submanifolds using the method of moving frames is due to Shiing-Shen Chern, and was presented in two papers by Chern and myself [37]–[38]. These two papers and indeed this monograph grew out of my work with Professor Chern during my 1985–

1986 sabbatical at Berkeley. I am very grateful to Professor Chern for many helpful discussions and insights.

I also want to thank several other mathematicians for their personal contributions. Katsumi Nomizu introduced me to Pinkall's work and Lie sphere geometry in 1982, and his seminar at Brown University has been the site of many enlightening discussions on the subject since that time. Thomas Banchoff introduced me to the cyclides of Dupin in the early seventies, when I was a graduate student, and he has provided me with several key insights over the years, particularly through his films. Patrick Ryan has contributed significantly to my understanding of this subject through many lectures and discussions. I also want to acknowledge helpful conversations and correspondence on various aspects of the subject with Steven Buyske, Sheila Carter, Leslie Coghlan, Josef Dorfmeister, Thomas Hawkins, Wu-Yi Hsiang, Nicolaas Kuiper, Martin Magid, Reiko Miyaoka, Ross Niebergall, Tetsuya Ozawa, Richard Palais, Ulrich Pinkall, Helmut Reckziegel, Chuu-Lian Terng, Gudlaugur Thorbergsson, and Alan West.

This book grew out of lectures given in the Brown University Differential Geometry Seminar in 1982–83 and subsequent lectures given to the Clavius Group during the summers of 1985–1989 at the University of Notre Dame, the University of California at Berkeley, Fairfield University and the Institute for Advanced Study. I want to thank my fellow members of the Clavius Group for their support of these lectures and many enlightening remarks. I also acknowledge with gratitude the hospitality of the institutions mentioned above.

I wish to thank my colleagues in the Department of Mathematics at the College of the Holy Cross, several of whom are my former teachers, for many insights and much encouragement over the years. I especially wish to mention my first teacher in linear algebra and real analysis, Leonard Sulski, who recently passed away after a courageous battle against leukemia. Professor Sulski was a superb, dedicated teacher, and a good and generous man. He will be missed by all who knew him.

While writing this book, I was supported by grants from the National Science Foundation (DMS-8907366 and DMS-9101961) and by a Faculty Fellowship from the College of the Holy Cross. This support is gratefully acknowledged.

I want to thank my three undergraduate research assistants from Holy Cross, Michele Intermont, Christopher Butler and Karen Purtell, who were also supported by the NSF. They worked through various versions of the manuscript and made many helpful comments. I also wish to thank the mathematics editorial department of Springer-Verlag for their timely professional help in preparing this manuscript for publication, and Kenneth Scott of Holy Cross for his assistance with the word-processing program.

Finally, I am most grateful to my wife, Patsy, and my sons, Tom, Mark, and Michael, for their patience, understanding and encouragement during this lengthy project.

August, 1991

Preface to the Second Edition

The most significant changes in the second edition are the following. First of all, this edition of the manuscript was prepared using the LaTeX document preparation system, and thus all of the numbers of the equations, theorems, etc., are different from the first edition.

I have added a new section Section 4.7 which describes the construction due to Ferus, Karcher, and Münzner [73] of isoparametric hypersurfaces with four principal curvatures, based on representations of Clifford algebras. Our treatment follows the original paper of Ferus, Karcher, and Münzner quite closely. I have also substantially revised the presentation of the invariance of tautness under Lie sphere transformations in Section 4.6, giving a different proof, due to J. C. Álvarez Paiva [2], who used functions whose level sets form a parabolic pencil of spheres rather than the usual distance functions to formulate tautness. This leads to a very natural proof of the Lie invariance of tautness.

Sections 5.2–5.4 regarding reducible Dupin hypersurfaces and the cyclides of Dupin have also been significantly revised, and 11 new figures illustrating the cyclides of Dupin have been added to Section 5.4. The introduction and several other places, for example Section 4.5, in the text where surveys of known results are given have all been updated to reflect the current state of research.

All 14 of the figures from the first edition were redone, and the second edition contains 14 additional figures. All of these figures were constructed by my colleague at Holy Cross, Andrew D. Hwang, using his ePiX program for constructing figures in the LaTeX picture environment. I am most grateful to Professor Hwang for the excellent quality of the figures, and for his time and effort in constructing them. The project page for the ePiX program is: http://math.holycross.edu/~ahwang/software/ePiX.html

In addition to the many mathematicians that I acknowledged in the preface to the first edition, I wish to thank Quo–Shin Chi and Gary Jensen, with whom I have collaborated on joint research over the past decade. This collaboration has been most stimulating and has led me to a deeper understanding of many aspects of this subject, in particular the method of moving frames and its applications to isoparametric and Dupin hypersurfaces.

I would also again like to thank the members of the Clavius Group for their encouragement and for their support of my lectures on Lie sphere geometry given at the University of Notre Dame in July, 2005 and the College of the Holy Cross in July, 2006. I also want to thank those institutions for their hospitality during the Clavius Group meetings.

While completing this second edition, I was supported by a grant from the National Science Foundation (DMS–0405529) and by a sabbatical leave from the College of the Holy Cross. This support is gratefully acknowledged. I also wish to thank my three recent undergraduate research assistants from Holy Cross, Ellen Gasparovic, Heather Johnson and Renee Laverdiere, who were also supported by my NSF grant, and who helped me in revising this book in various ways.

I am also grateful to Ann Kostant and her staff at Springer for their support and encouragement to complete a second edition of the book.

Finally, as with the first edition, I wish to thank my wife, Patsy, and my sons, Tom, Mark, and Michael, for their warm encouragement and support of all of my scholarly efforts, in particular, the second edition of this book.

College of the Holy Cross   *Thomas E. Cecil*
Worcester, Massachusetts
August, 2007

Contents

1

Introduction

Lie [104] introduced the geometry of oriented spheres in his dissertation, published as a paper in *Mathematische Annalen* in 1872. Sphere geometry was also prominent in his study of contact transformations (Lie–Scheffers [105]) and in Volume III of Blaschke's book on differential geometry published in 1929. In recent years, Lie sphere geometry has become a valuable tool in the study of Dupin submanifolds, beginning with Pinkall's [146] dissertation in 1981. In this introduction, we will outline the contents of the book and mention some related results.

Lie established a bijective correspondence between the set of all *Lie spheres*, i.e., oriented hyperspheres, oriented hyperplanes and point spheres in $\mathbf{R}^n \cup \{\infty\}$, and the set of all points on the quadric hypersurface Q^{n+1} in real projective space $\mathbf{P}^{n+2}$ given by the equation $\langle x, x \rangle = 0$, where $\langle , \rangle$ is an indefinite scalar product with signature $(n + 1, 2)$ on $\mathbf{R}^{n+3}$. This *Lie quadric* Q^{n+1} contains projective lines but no linear subspaces of higher dimension. The one-parameter family of Lie spheres in $\mathbf{R}^n \cup \{\infty\}$ corresponding to the points on a line on Q^{n+1} is called a *parabolic pencil* of spheres. It consists of all Lie spheres in oriented contact at a certain contact element on $\mathbf{R}^n \cup \{\infty\}$. In this way, Lie also established a bijective correspondence between the manifold of contact elements on $\mathbf{R}^n \cup \{\infty\}$ and the manifold Λ^{2n-1} of projective lines on the Lie quadric Q^{n+1}. The details of these considerations are given in Chapter 2.

A *Lie sphere transformation* is a projective transformation of $\mathbf{P}^{n+2}$ which maps the Lie quadric Q^{n+1} to itself. In terms of the geometry of $\mathbf{R}^n$, a Lie sphere transformation maps Lie spheres to Lie spheres. Furthermore, since a projective transformation maps lines to lines, a Lie sphere transformation preserves oriented contact of Lie spheres in $\mathbf{R}^n$. Lie proved the so-called "fundamental theorem of Lie sphere geometry" in the case $n = 3$, and Pinkall [150] generalized this to higher dimensions. This theorem states that a line-preserving diffeomorphism of the Lie quadric Q^{n+1} is the restriction to Q^{n+1} of a projective transformation of $\mathbf{P}^{n+2}$. In other words, a transformation on the space of Lie spheres that preserves oriented contact of spheres is a Lie sphere transformation. One can show that a Lie sphere transformation is induced by an orthogonal transformation of $\mathbf{R}^{n+3}$ endowed with the metric $\langle , \rangle$. Thus, the group G of Lie sphere transformations is isomorphic to the quotient group

$O(n + 1, 2)/\{\pm I\}$. By the theorem of Cartan and Dieudonné, the orthogonal group $O(n + 1, 2)$ is generated by inversions in hyperplanes, and therefore so is G. Any Möbius (conformal) transformation of $\mathbf{R}^n \cup \{\infty\}$ induces a Lie sphere transformation, and the Möbius group is precisely the subgroup of Lie sphere transformations that map point spheres to point spheres. In Chapter 3, we prove these results and give a geometric description of inversions. We also discuss the sphere geometries of Laguerre and Möbius. These, as well as the usual Euclidean, spherical and hyperbolic geometries, are subgeometries of Lie sphere geometry.

The manifold Λ^{2n-1} of projective lines on the quadric Q^{n+1} has a contact structure, i.e., a 1-form ω such that $\omega \wedge (d\omega)^{n-1}$ does not vanish on Λ^{2n-1}. The condition $\omega = 0$ defines a codimension one distribution D on Λ^{2n-1} which has integral submanifolds of dimension $n - 1$ but none of higher dimension. An immersion $\lambda : M^{n-1} \to \Lambda^{2n-1}$ of an $(n - 1)$-dimensional manifold M^{n-1} such that $\lambda^* \omega = 0$ is called a *Legendre submanifold*. If α is a Lie sphere transformation, then α maps lines to lines, and the map $\mu = \alpha\lambda$ is also a Legendre submanifold. The submanifolds λ and μ are said to be *Lie equivalent*. Legendre submanifolds are studied in detail in Chapter 4.

A hypersurface M in $\mathbf{R}^n$ naturally induces a Legendre submanifold. More generally, an immersed submanifold V of codimension greater than one in $\mathbf{R}^n$ induces a Legendre submanifold whose domain is the unit normal bundle B^{n-1} of V in $\mathbf{R}^n$. Thus, Lie sphere geometry can be used to study any problem concerning submanifolds of $\mathbf{R}^n$, or more generally of the sphere S^n or hyperbolic space H^n. Of course, Lie sphere geometry is particularly well-suited for the study of problems that deal with spheres in some way. A large class of such problems are those involving the principal curvatures of a submanifold, since each principal curvature gives rise to a corresponding curvature sphere.

Let M be a hypersurface in a real space-form $\mathbf{R}^n$, S^n or H^n. The eigenvalues of the shape operator (second fundamental form) A of M are called *principal curvatures*, and their corresponding eigenspaces are called *principal spaces*. A submanifold S of M is called a *curvature surface* if at each point of S, the tangent space T_xS is a principal space. This generalizes the classical notion of a line of curvature of a surface in $\mathbf{R}^3$. Curvature surfaces are abundant, for there always exists an open dense subset Ω of M on which the multiplicities of the principal curvatures are locally constant (see Reckziegel [157]–[158]). If a principal curvature κ has constant multiplicity m in some open set $U \subset M$, then the corresponding distribution of principal spaces is an m-dimensional foliation, and the leaves of this *principal foliation* are curvature surfaces. Furthermore, if the multiplicity m of κ is greater than one, then κ is constant along each leaf of this principal foliation. This is not true, in general, if $m = 1$. A hypersurface M is said to be *Dupin* if along each curvature surface, the corresponding principal curvature is constant. A Dupin hypersurface is said to be *proper Dupin* if each principal curvature has constant multiplicity on M, i.e., the number of distinct principal curvatures is constant on M. An example of a proper Dupin hypersurface in $\mathbf{R}^3$ is a torus of revolution T^2. There exist many examples of Dupin hypersurfaces that are not proper Dupin, e.g., a tube M^3 in $\mathbf{R}^4$ of constant radius over a torus of revolution $T^2 \subset \mathbf{R}^3 \subset \mathbf{R}^4$ (see Section 5.2 or Pinkall [150]).

The notion of Dupin can be generalized to submanifolds of higher codimension in $\mathbf{R}^n$ or even to the larger class of Legendre submanifolds. Moreover, the Dupin and proper Dupin properties are easily seen to be invariant under Lie sphere transformations. This makes Lie sphere geometry a particularly effective setting for the study of Dupin submanifolds.

Noncompact proper Dupin hypersurfaces in real space-forms are plentiful. Pinkall [150] introduced four constructions for obtaining a proper Dupin hypersurface W in $\mathbf{R}^{n+m}$ from a proper Dupin hypersurface M in $\mathbf{R}^n$. These constructions involve building tubes, cylinders, cones and surfaces of revolution from M, and they are discussed in detail in Sections 5.1–5.3. Using these constructions, Pinkall was able to construct a proper Dupin hypersurface in Euclidean space with an arbitrary number of distinct principal curvatures with any given multiplicities. In general, these proper Dupin hypersurfaces cannot be extended to compact Dupin hypersurfaces without losing the property that the number of distinct principal curvatures is constant. Proper Dupin hypersurfaces that are locally Lie equivalent to the end product of one of Pinkall's constructions are said to be *reducible*.

Compact proper Dupin submanifolds are much more rare. Important examples are obtained through a consideration of isoparametric hypersurfaces. An immersed hypersurface M in a real space-form, $\mathbf{R}^n$, S^n or H^n, is said to be *isoparametric* if it has constant principal curvatures. An isoparametric hypersurface M in $\mathbf{R}^n$ can have at most two distinct principal curvatures, and M must be an open subset of a hyperplane, hypersphere or a spherical cylinder $S^k \times \mathbf{R}^{n-k-1}$. This was shown by Levi–Civita [101] for $n = 3$ and by B. Segre [169] for arbitrary n.

E. Cartan [16]–[19] began the study of isoparametric hypersurfaces in the other space-forms in a series of four papers in the 1930s. In hyperbolic space H^n, he showed that an isoparametric hypersurface can have at most two distinct principal curvatures, and it is either totally umbilic or else a standard product $S^k \times H^{n-k-1}$ in H^n (see also Ryan [164, pp. 252–253]).

In the sphere S^n, however, Cartan showed that there are many more possibilities. He found examples of isoparametric hypersurfaces in S^n with $1, 2, 3$ or 4 distinct principal curvatures, and he classified compact, connected isoparametric hypersurfaces with $g \leq 3$ principal curvatures as follows. If $g = 1$, then the isoparametric hypersurface M is totally umbilic, and it must be a great or small sphere. If $g = 2$, then M must be a standard product of two spheres,

$$S^k(r) \times S^{n-k-1}(s) \subset S^n, \quad r^2 + s^2 = 1.$$

In the case $g = 3$, Cartan [17] showed that all the principal curvatures must have the same multiplicity $m = 1, 2, 4$ or 8, and the isoparametric hypersurface must be a tube of constant radius over a standard embedding of a projective plane $\mathbf{F}P^2$ into S^{3m+1} (see, for example, Cecil–Ryan [52, pp. 296–299]), where $\mathbf{F}$ is the division algebra $\mathbf{R}$, $\mathbf{C}$, $\mathbf{H}$ (quaternions), $\mathbf{O}$ (Cayley numbers), for $m = 1, 2, 4, 8$, respectively. Thus, up to congruence, there is only one such family for each value of m.

Cartan's theory was further developed by Nomizu [135]–[136], Takagi and Takahashi [180], Ozeki and Takeuchi [143], and most extensively by Münzner [123], who showed that the number g of distinct principal curvatures of an isoparametric hypersurface must be 1, 2, 3, 4, or 6. (See also Cecil–Ryan [52, Chapter 3] or the survey article by Thorbergsson [192].)

In the case of an isoparametric hypersurface with four principal curvatures, Münzner proved that the principal curvatures can have at most two distinct multiplicities m_1, m_2. Ferus, Karcher, and Münzner [73] then used representations of Clifford algebras to construct for any positive integer m_1 an infinite series of isoparametric hypersurfaces with four principal curvatures having respective multiplicities (m_1, m_2), where m_2 is nondecreasing and unbounded in each series. This class of *FKM-type* isoparametric hypersurfaces (described in Section 4.7) contains all known examples of isoparametric hypersurfaces with four principal curvatures with the exception of two homogeneous examples, having multiplicities $(2, 2)$ and $(4, 5)$.

Stolz [177] then proved that the multiplicities of the principal curvatures of an isoparametric hypersurface with four principal curvatures must be the same as those in the known examples of FKM-type or the two homogeneous exceptions. Cecil, Chi and Jensen [40] next showed that an isoparametric hypersurface with four principal curvatures must be of FKM-type, if the multiplicities satisfy $m_2 \geq 2m_1 - 1$ (a different proof of this result, using isoparametric triple systems, was given by Immervoll [92]). Taken together with known results of Takagi [179] for $m_1 = 1$ and Ozeki and Takeuchi [143] for $m_1 = 2$, this handles all possible pairs of multiplicities except for four cases, for which the classification problem remains open.

The case of isoparametric hypersurfaces with $g = 6$ distinct principal curvatures also remains open. In that case, there exists one homogeneous family with six principal curvatures of multiplicity one in S^7, and one homogeneous family with six principal curvatures of multiplicity two in S^{13} (see Miyaoka [118] for a description). Up to congruence, these are the only known examples. Münzner showed that for an isoparametric hypersurface with six principal curvatures, all of the principal curvatures must have the same multiplicity m. Abresch [1] then showed that m must be 1 or 2. In the case of multiplicity $m = 1$, Dorfmeister and Neher [62] showed in 1985 that an isoparametric hypersurface must be homogeneous (see also Miyaoka [119]–[120] for an alternative proof of this result). The case $m = 2$ remains open, although it is generally thought that the hypersurface must be homogeneous in that case also.

There is also an extensive theory of isoparametric submanifolds of codimension greater than one in the sphere, due primarily to Terng [183]–[187], Hsiang, Palais and Terng [91], and Carter and West [26]–[28], [199]. (See also Harle [82] and Strübing [178].) A connected, complete submanifold V in a real space-form is said to be *isoparametric* if it has flat normal bundle and if for any parallel section of the unit normal bundle $\eta : V \rightarrow B^{n-1}$, the principal curvatures of A_η are constant. After considerable development of the theory, Thorbergsson [191] showed that all isoparametric submanifolds of codimension greater than one in S^n are homogeneous, and are thus principal orbits of isotropy representations of symmetric spaces, also known as generalized flag manifolds or standard embeddings of R-spaces. See Olmos [139] and Heintze–Liu [87] for alternate proofs of Thorbergsson's result, and the papers of Bott and Samelson [12], Takeuchi and Kobayashi [182] and Hahn [81] for more on generalized flag manifolds and R-spaces. The paper of Heintze, Olmos and

Thorbergsson [88], and the books by Palais and Terng [145], and Berndt, Console and Olmos [8] contain further results in this area.

For a generalization of the theory of isoparametric submanifolds to submanifolds of hyperbolic space, see Wu [201] and Zhao [203]. For isoparametric hypersurfaces in Lorentz spaces, see Nomizu [137], Magid [108] and Hahn [80]. For isoparametric and Dupin hypersurfaces in affine differential geometry, see Niebergall and Ryan [128]–[131]. For related notions for submanifolds of symmetric spaces, see Terng and Thorbergsson [188], and West [200].

Terng [186] introduced a theory of isoparametric submanifolds in infinite-dimensional Hilbert spaces. Pinkall and Thorbergsson [154] then provided more examples of such submanifolds, and Heintze and Liu [87] proved a homogeneity result for these submanifolds.

Thorbergsson [190] showed that the restriction $g = 1, 2, 3, 4$ or 6 on the number of distinct principal curvatures also holds for a connected, compact proper Dupin hypersurface M embedded in $S^n \subset \mathbf{R}^{n+1}$. He first showed that M must be *taut*, i.e., every nondegenerate distance function $L_p(x) = |p - x|^2, p \in \mathbf{R}^{n+1}$, has the minimum number of critical points required by the Morse inequalities on M. Using tautness, he then showed that M divides S^n into two ball bundles over the first focal submanifolds on either side of M. This topological situation is all that is required for Münzner's proof of the restriction on g. Münzner's argument also produces certain restrictions on the cohomology and the homotopy groups of isoparametric hypersurfaces. These restrictions necessarily apply to compact proper Dupin hypersurfaces by Thorbergsson's result. Grove and Halperin [79] later found more topological similarities between these two classes of hypersurfaces. Furthermore, the results of Stolz [177] on the possible multiplicities of the principal curvatures actually only require the assumption that M is a compact proper Dupin hypersurface, and not that the hypersurface is isoparametric.

The close relationship between these two classes of hypersurfaces led to the widely held conjecture that every compact proper Dupin hypersurface $M \subset S^n$ is equivalent by a Lie sphere transformation to an isoparametric hypersurface (see [52, p. 184]). The conjecture is obviously true for $g = 1$, in which case M must be a hypersphere in S^n, and so M itself is isoparametric. In 1978, Cecil and Ryan [47] showed that if $g = 2$, then M must be a cyclide of Dupin, and it is therefore Möbius equivalent to an isoparametric hypersurface. Then in 1984, Miyaoka [111] showed that the conjecture holds for $g = 3$, although it is not true that M must be Möbius equivalent to an isoparametric hypersurface. Thus, as g increases, the group needed to obtain equivalence with an isoparametric hypersurface gets progressively larger. The case $g = 4$ resisted all attempts at solution for several years and finally in 1988, counterexamples to the conjecture were discovered independently by Pinkall and Thorbergsson [152] and by Miyaoka and Ozawa [121]. The latter method also yields counterexamples to the conjecture with $g = 6$ principal curvatures. In both constructions, a fundamental Lie invariant, the Lie curvature (see Section 4.5), was used to show that the examples are not Lie equivalent to an isoparametric hypersurface. Specifically, if M is a proper Dupin hypersurface with four principal curvatures, then the *Lie curvature* Ψ is defined to be the cross-ratio of these principal curvatures. Viewed in the context of projective geometry, Ψ is the cross-ratio of the four points along a projective line on Q^{n+1} corresponding to the four curvature spheres of M. Since a Lie sphere transformation maps curvature spheres to curvature spheres and preserves cross-ratios, the Lie curvature is invariant under Lie sphere transformations. From the work of Münzner, it is easy to show that Ψ has the constant value $1/2$ (when the principal curvatures are appropriately ordered) on an isoparametric hypersurface with four principal curvatures. For the counterexamples to the conjecture, it was shown that $\Psi \neq 1/2$ at some points, and therefore the examples cannot be Lie equivalent to an isoparametric hypersurface. These examples are presented in detail in Section 4.8.

In Section 4.6, we show that tautness is invariant under Lie sphere transformations. A proof of this result was first given in [37]. However, the proof that we give here is due to Álvarez Paiva [2], who used functions whose level sets form a parabolic pencil of spheres rather than the usual distance functions to formulate tautness. From this point of view, the Lie invariance of tautness is seen quite easily. Tautness is closely related to the Dupin condition. Every taut submanifold of a real space-form is Dupin, although not necessarily proper Dupin. (Pinkall [151], and Miyaoka [112], independently, for hypersurfaces.) Conversely, Thorbergsson [190] proved that a compact proper Dupin hypersurface embedded in S^n is taut. Pinkall [151] then extended this result to compact submanifolds of higher codimension for which the number of distinct principal curvatures is constant on the unit normal bundle. An open question is whether the Dupin condition implies tautness without this assumption. A key fact in Thorbergsson's proof is that in the proper Dupin case, all the curvature surfaces are spheres. In the nonproper Dupin case, the work of Ozawa [142] implies that some of the curvature surfaces are not spheres.

A submanifold M of $\mathbf{R}^n$ is said to be *totally focal* if every distance function L_p is either nondegenerate or has only degenerate focal points on M. For the relationship between isoparametric and totally focal submanifolds, see the papers of Carter and West [22, 23, 24, 28].

Chapter 5 is devoted primarily to the local classification of proper Dupin hypersurfaces in certain specific cases. These results were obtained by Lie sphere geometric techniques and have not been proved by standard Euclidean methods. In Section 5.4, we give Pinkall's [150] local classification of proper Dupin submanifolds with two distinct principal curvatures. These are known as the *cyclides of Dupin*. This is followed by a classification of the cyclides of Dupin up to Möbius (conformal) transformations that can be derived from the Lie sphere geometric classification. Finally, in Section 5.7, we present the classification of proper Dupin hypersurfaces in $\mathbf{R}^4$ with three distinct principal curvatures. This was first obtained by Pinkall [146], [149], although the treatment here is due to Cecil and Chern [38]. In the process, we develop the method of moving Lie frames which can be applied to the general study of Legendre submanifolds. This approach has been applied sucessfully by Niebergall [126]–[127], Cecil and Jensen [44]–[45] and by Cecil, Chi and Jensen [41] to obtain local classifications of higher-dimensional Dupin hypersurfaces.

In particular, Cecil and Jensen [44] proved that for a connected irreducible proper Dupin hypersurface with three principal curvatures, all of the multiplicities must be equal, and the hypersurface must be Lie equivalent to an open subset of an isoparametric hypersurface with three principal curvatures.

In the case of an isoparametric hypersurface with four principal curvatures, Münzner showed that the multiplicities of the principal curvatures must satisfy $m_1 = m_2$, $m_3 = m_4$, when the principal curvatures are appropriately ordered, and that the Lie curvature Ψ must have the constant value -1 (when the principal curvatures are ordered in this way). In [45, pp. 3–4], it was conjectured that an irreducible connected proper Dupin hypersurface M in S^n with four principal curvatures having multiplicities satisfying $m_1 = m_2$, $m_3 = m_4$ and constant Lie curvature must be Lie equivalent to an open subset of an isoparametric hypersurface in S^n. In that same paper, the conjecture was shown to be true in the case where all the multiplicities are equal to one. Later in [41], this was generalized to show that the conjecture is true if the multiplicities satisfy $m_1 = m_2 \geq 1$, $m_3 = m_4 = 1$, and the Lie curvature has the constant value $\Psi = -1$. The conjecture in its full generality remains open, although it is still thought to be true.

There are many aspects of Lie sphere geometry that are not covered in detail here. In particular, Blaschke [10] gives a more thorough treatment of the sphere geometries of Laguerre and Möbius and the "line-sphere transformation" of Lie (see Blaschke [10, Section 54] and Klein [94, Section 70]). The line-sphere transformation is discussed in a more modern setting by Fillmore [75], who also treats the relationship between Lie sphere geometry and complex line geometry. A modern treatment of Möbius differential geometry is given in the book by Hertrich–Jeromin [89] or the papers of C.-P. Wang [194]–[196]. For submanifold theory in Laguerre geometry, see Blaschke [10], Musso and Nicolodi [124]–[125], Li [102], or Li and Wang [103].

In this book, we concentrate on submanifolds of dimension greater than one in real space forms. The papers of Sasaki and Suguri [167] and Pinkall [147] treat curve theory in Lie sphere geometry.

Other work involving submanifolds in the context of Lie sphere geometry includes two papers of Miyaoka [115]–[116] that extend some of the key ideas of the Lie sphere geometric approach to the study of contact structures and conformal structures on more general manifolds. Ferapontov [70]–[72] studied the relationship between Dupin and isoparametric hypersurfaces and integrable Hamiltonian systems of hydrodynamic type. Riveros and Tenenblat [160]–[161] used higher-dimensional Laplace invariants to study four-dimensional Dupin hypersurfaces. Finally, the history and significance of Lie's early work on sphere geometry and contact transformations is discussed in the papers of Hawkins [83] and Rowe [162].

All manifolds and maps are assumed to be smooth unless explicitly stated otherwise. Notation generally follows Kobayashi and Nomizu [95] and Cecil and Ryan [52]. Theorems, equations, remarks, etc., are numbered within each chapter. If a theorem from a different chapter is cited, then the chapter is listed along with the number of the theorem.

2

Lie Sphere Geometry

In this chapter, we give Lie's construction of the space of spheres and define the important notions of oriented contact and parabolic pencils of spheres. This leads ultimately to a bijective correspondence between the manifold of contact elements on the sphere S^n and the manifold Λ^{2n-1} of projective lines on the Lie quadric.

2.1 Preliminaries

Before constructing the space of spheres, we begin with some preliminary remarks on indefinite scalar product spaces and projective geometry. Finite-dimensional indefinite scalar product spaces play a crucial role in Lie sphere geometry. The fundamental result from linear algebra concerns the rank and signature of a bilinear form (see, for example, Nomizu [133, p. 108], Artin [4, Chapter 3], or O'Neill [141, pp. 46–53]).

Theorem 2.1. *Suppose that* $(,)$ *is a bilinear form on a real vector space* V *of dimension* n. *Then there exists a basis* $\{e_1, \ldots, e_n\}$ *of* V *such that*

1. $(e_i, e_j) = 0$ *for* $i \neq j$,
2. $(e_i, e_i) = 1$ *for* $1 \leq i \leq p$,
3. $(e_j, e_j) = -1$ *for* $p + 1 \leq j \leq r$,
4. $(e_k, e_k) = 0$ *for* $r + 1 \leq k \leq n$.

The numbers r and p are determined solely by the bilinear form; r is called the *rank*, $r - p$ is called the *index*, and the ordered pair $(p, r - p)$ is called the *signature* of the bilinear form. The theorem shows that any two spaces of the same dimension with bilinear forms of the same signature are isometrically isomorphic. A *scalar product* is a nondegenerate bilinear form, i.e., a form with rank equal to the dimension of V. For the sake of brevity, we will often refer to a scalar product as a "metric." Usually, we will be dealing with the scalar product space $\mathbf{R}_k^n$ with signature $(n - k, k)$ for $k = 0, 1$ or 2. However, at times we will consider subspaces of $\mathbf{R}_k^n$ on which the bilinear form is degenerate. When dealing with low-dimensional spaces, we will often indicate the signature with a series of plus and minus signs and zeroes where appropriate. For example, the signature of $\mathbf{R}_1^3$ may be written $(+ + -)$ instead of $(2, 1)$. If the bilinear form is nondegenerate, a basis with the properties listed in Theorem 2.1 is called an *orthonormal basis* for V with respect to the bilinear form.

A second useful result concerning scalar products is the following. Here $U^\perp$ denotes the orthogonal complement of the space U with respect to the given scalar product. (See Artin [4, p. 117] or O'Neill [141, p. 49].)

Theorem 2.2. *Suppose that $(,)$ is a scalar product on a finite-dimensional real vector space V and that U is a subspace of V.*

(a) *Then $U^{\perp\perp} = U$ and $\dim U + \dim U^\perp = \dim V$.*
(b) *The form $(,)$ is nondegenerate on U if and only if it is nondegenerate on $U^\perp$. If the form is nondegenerate on U, then V is the direct sum of U and $U^\perp$.*
(c) *If V is the orthogonal direct sum of two spaces U and W, then the form is nondegenerate on U and W, and $W = U^\perp$.*

Let (x, y) be the indefinite scalar product on the Lorentz space $\mathbf{R}_1^{n+1}$ defined by

$$(x, y) = -x_1 y_1 + x_2 y_2 + \cdots + x_{n+1} y_{n+1}, \tag{2.1}$$

where $x = (x_1, \ldots, x_{n+1})$ and $y = (y_1, \ldots, y_{n+1})$. We will call this scalar product the *Lorentz metric*. A vector x is said to be *spacelike*, *timelike* or *lightlike*, respectively, depending on whether (x, x) is positive, negative or zero. We will use this terminology even when we are using a metric of different signature. In Lorentz space, the set of all lightlike vectors, given by the equation

$$x_1^2 = x_2^2 + \cdots + x_{n+1}^2, \tag{2.2}$$

forms a cone of revolution, called the *light cone*. Lightlike vectors are often called *isotropic* in the literature, and the cone is called the *isotropy cone*. Timelike vectors are "inside the cone" and spacelike vectors are "outside the cone."

If x is a nonzero vector in $\mathbf{R}_1^{n+1}$, let $x^\perp$ denote the orthogonal complement of x with respect to the Lorentz metric. If x is timelike, then the metric restricts to a positive definite form on $x^\perp$, and $x^\perp$ intersects the light cone only at the origin. If x is spacelike, then the metric has signature $(n - 1, 1)$ on $x^\perp$, and $x^\perp$ intersects the cone in a cone of one less dimension. If x is lightlike, then $x^\perp$ is tangent to the cone along the line through the origin determined by x. The metric has signature $(n - 1, 0)$ on this n-dimensional plane.

The true setting of Lie sphere geometry is real projective space $\mathbf{P}^n$, so we now briefly review some important concepts from projective geometry. We define an equivalence relation on $\mathbf{R}^{n+1} - \{0\}$ by setting $x \simeq y$ if $x = ty$ for some nonzero real number t. We denote the equivalence class determined by a vector x by $[x]$. Projective space $\mathbf{P}^n$ is the set of such equivalence classes, and it can naturally be identified with the space of all lines through the origin in $\mathbf{R}^{n+1}$. The rectangular coordinates $(x_1, \ldots, x_{n+1})$ are called *homogeneous coordinates* of the point $[x]$, and they are only determined up to a nonzero scalar multiple. The affine space $\mathbf{R}^n$ can be embedded in $\mathbf{P}^n$ as the complement of the hyperplane $(x_1 = 0)$ at infinity by the map $\phi : \mathbf{R}^n \to \mathbf{P}^n$ given by $\phi(u) = [(1, u)]$. A scalar product on $\mathbf{R}^{n+1}$, such as the Lorentz metric, determines a polar relationship between points and hyperplanes in $\mathbf{P}^n$. We will also use the notation $x^{\perp}$ to denote the polar hyperplane of $[x]$ in $\mathbf{P}^n$, and we will call $[x]$ the *pole* of $x^{\perp}$.

If x is a lightlike vector in $\mathbf{R}_1^{n+1}$, then $[x]$ can be represented by a vector of the form $(1, u)$ for $u \in \mathbf{R}^n$. Then the equation $(x, x) = 0$ for the light cone becomes $u \cdot u = 1$ (Euclidean dot product), i.e., the equation for the unit sphere in $\mathbf{R}^n$. Hence, the set of points in $\mathbf{P}^n$ determined by lightlike vectors in $\mathbf{R}_1^{n+1}$ is naturally diffeomorphic to the sphere S^{n-1}.

2.2 Möbius Geometry of Unoriented Spheres

As a first step toward Lie sphere geometry, we recall the geometry of unoriented spheres in $\mathbf{R}^n$ known as "Möbius" or "conformal" geometry. We will always assume that $n \geq 2$. In this section, we will only consider spheres and planes of codimension one, and we will often omit the prefix "hyper."

We denote the Euclidean dot product of two vectors u and v in $\mathbf{R}^n$ by $u \cdot v$. We first consider stereographic projection $\sigma : \mathbf{R}^n \to S^n - \{P\}$, where S^n is the unit sphere in $\mathbf{R}^{n+1}$ given by $y \cdot y = 1$, and $P = (-1, 0, \ldots, 0)$ is the south pole of S^n. (See Figure 2.1.) The well-known formula for $\sigma(u)$ is

$$\sigma(u) = \left(\frac{1 - u \cdot u}{1 + u \cdot u}, \frac{2u}{1 + u \cdot u} \right).$$

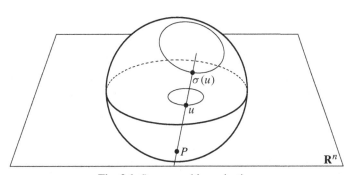

Fig. 2.1. Stereographic projection.

We next embed $\mathbf{R}^{n+1}$ into $\mathbf{P}^{n+1}$ by the embedding ϕ mentioned in the previous section. Thus, we have the map $\phi\sigma : \mathbf{R}^n \to \mathbf{P}^{n+1}$ given by

$$\phi\sigma(u) = \left[\left(1, \frac{1 - u \cdot u}{1 + u \cdot u}, \frac{2u}{1 + u \cdot u} \right) \right] = \left[\left(\frac{1 + u \cdot u}{2}, \frac{1 - u \cdot u}{2}, u \right) \right]. \quad (2.3)$$

Let $(z_1, \ldots, z_{n+2})$ be homogeneous coordinates on $\mathbf{P}^{n+1}$ and $(\,,\,)$ the Lorentz metric on the space $\mathbf{R}_1^{n+2}$. Then $\phi\sigma(\mathbf{R}^n)$ is just the set of points in $\mathbf{P}^{n+1}$ lying on the n-sphere Σ given by the equation $(z, z) = 0$, with the exception of the *improper point* $[(1, -1, 0, \ldots, 0)]$ corresponding to the south pole P. We will refer to the points in Σ other than $[(1, -1, 0, \ldots, 0)]$ as *proper points*, and will call Σ the *Möbius sphere* or *Möbius space*. At times, it is easier to simply begin with S^n rather than $\mathbf{R}^n$ and thus avoid the need for the map σ and the special point P. However, there are also advantages for beginning in $\mathbf{R}^n$.

The basic framework for the Möbius geometry of unoriented spheres is as follows. Suppose that ξ is a spacelike vector in $\mathbf{R}_1^{n+2}$. Then the polar hyperplane $\xi^\perp$ to $[\xi]$ in $\mathbf{P}^{n+1}$ intersects the sphere Σ in an $(n-1)$-sphere S^{n-1} (see Figure 2.2). S^{n-1} is the image under $\phi\sigma$ of an $(n-1)$-sphere in $\mathbf{R}^n$, unless it contains the improper point $[(1, -1, 0, \ldots, 0)]$, in which case it is the image under $\phi\sigma$ of a hyperplane in $\mathbf{R}^n$. Hence, we have a bijective correspondence between the set of all spacelike points in $\mathbf{P}^{n+1}$ and the set of all hyperspheres and hyperplanes in $\mathbf{R}^n$.

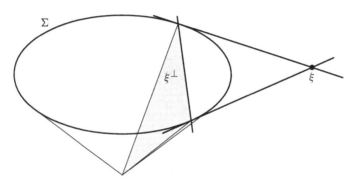

Fig. 2.2. Intersection of Σ with $\xi^\perp$.

It is often useful to have specific formulas for this correspondence. Consider the sphere in $\mathbf{R}^n$ with center p and radius $r > 0$ given by the equation

$$(u - p) \cdot (u - p) = r^2. \tag{2.4}$$

We wish to translate this into an equation involving the Lorentz metric and the corresponding polarity relationship on $\mathbf{P}^{n+1}$. A direct calculation shows that equation (2.4) is equivalent to the equation

$$(\xi, \phi\sigma(u)) = 0, \tag{2.5}$$

where ξ is the spacelike vector,

$$\xi = \left(\frac{1 + p \cdot p - r^2}{2}, \frac{1 - p \cdot p + r^2}{2}, p \right), \tag{2.6}$$

and $\phi\sigma(u)$ is given by equation (2.3). Thus, the point u is on the sphere given by equation (2.4) if and only if $\phi\sigma(u)$ lies on the polar hyperplane of $[\xi]$. Note that the first two coordinates of ξ satisfy $\xi_1 + \xi_2 = 1$, and that $(\xi, \xi) = r^2$. Although ξ is only determined up to a nonzero scalar multiple, we can conclude that $\eta_1 + \eta_2$ is not zero for any $\eta \simeq \xi$.

Conversely, given a spacelike point $[z]$ with $z_1 + z_2$ nonzero, we can determine the corresponding sphere in $\mathbf{R}^n$ as follows. Let $\xi = z/(z_1 + z_2)$ so that $\xi_1 + \xi_2 = 1$. Then from equation (2.6), the center of the corresponding sphere is the point $p = (\xi_3, \ldots, \xi_{n+2})$, and the radius is the square root of (ξ, ξ).

Next suppose that η is a spacelike vector with $\eta_1 + \eta_2 = 0$. Then

$$(\eta, (1, -1, 0, \ldots, 0)) = 0.$$

Thus, the improper point $\phi(P)$ lies on the polar hyperplane of $[\eta]$, and the point $[\eta]$ corresponds to a hyperplane in $\mathbf{R}^n$. Again we can find an explicit correspondence. Consider the hyperplane in $\mathbf{R}^n$ given by the equation

$$u \cdot N = h, \quad |N| = 1. \tag{2.7}$$

A direct calculation shows that (2.7) is equivalent to the equation

$$(\eta, \phi\sigma(u)) = 0, \quad \text{where } \eta = (h, -h, N). \tag{2.8}$$

Thus, the hyperplane (2.7) is represented in the polarity relationship by $[\eta]$. Conversely, let z be a spacelike point with $z_1 + z_2 = 0$. Then $(z, z) = v \cdot v$, where $v = (z_3, \ldots, z_{n+2})$. Let $\eta = z/|v|$. Then η has the form (2.8) and $[z]$ corresponds to the hyperplane (2.7). Thus we have explicit formulas for the bijective correspondence between the set of spacelike points in $\mathbf{P}^{n+1}$ and the set of hyperspheres and hyperplanes in $\mathbf{R}^n$.

Of course, the fundamental invariant of Möbius geometry is the angle. The study of angles in this setting is quite natural, since orthogonality between spheres and planes in $\mathbf{R}^n$ can be expressed in terms of the Lorentz metric. Let S_1 and S_2 denote the spheres in $\mathbf{R}^n$ with respective centers p_1 and p_2 and respective radii r_1 and r_2. By the Pythagorean theorem, the two spheres intersect orthogonally (see Figure 2.3) if and only if

$$|p_1 - p_2|^2 = r_1^2 + r_2^2. \tag{2.9}$$

If these spheres correspond by equation (2.6) to the projective points $[\xi_1]$ and $[\xi_2]$, respectively, then a calculation shows that equation (2.9) is equivalent to the condition

$$(\xi_1, \xi_2) = 0. \tag{2.10}$$

A hyperplane π in $\mathbf{R}^n$ is orthogonal to a hypersphere S precisely when π passes through the center of S. If S has center p and radius r, and π is given by the equation $u \cdot N = h$, then the condition for orthogonality is just $p \cdot N = h$. If S corresponds to $[\xi]$ as in (2.6) and π corresponds to $[\eta]$ as in (2.8), then this equation for orthogonality is equivalent to $(\xi, \eta) = 0$. Finally, if two planes π_1 and π_2 are represented by $[\eta_1]$

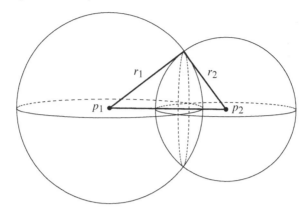

Fig. 2.3. Orthogonal spheres.

and $[\eta_2]$ as in (2.8), then the orthogonality condition $N_1 \cdot N_2 = 0$ is equivalent to the equation $(\eta_1, \eta_2) = 0$.

A *Möbius transformation* is a projective transformation of $\mathbf{P}^{n+1}$ which preserves the condition $(\eta, \eta) = 0$. By Theorem 3.1 of Chapter 3, p. 26, a Möbius transformation also preserves the relationship $(\eta, \xi) = 0$, and it maps spacelike points to spacelike points. Thus it preserves orthogonality (and hence angles) between spheres and planes in $\mathbf{R}^n$. In the next chapter, we will see that the group of Möbius transformations is isomorphic to $O(n + 1, 1)/\{\pm I\}$, where $O(n + 1, 1)$ is the group of orthogonal transformations of the Lorentz space $\mathbf{R}_1^{n+2}$.

Note that a Möbius transformation takes lightlike vectors to lightlike vectors, and so it induces a conformal diffeomorphism of the sphere Σ onto itself. It is well known that the group of conformal diffeomorphisms of the sphere is precisely the Möbius group.

2.3 Lie Geometry of Oriented Spheres

We now turn to the construction of Lie's geometry of oriented spheres and planes in $\mathbf{R}^n$. Let W^{n+1} be the set of vectors in $\mathbf{R}_1^{n+2}$ satisfying $(\zeta, \zeta) = 1$. This is a hyperboloid of revolution of one sheet in $\mathbf{R}_1^{n+2}$. If α is a spacelike point in $\mathbf{P}^{n+1}$, then there are precisely two vectors $\pm \zeta$ in W^{n+1} with $\alpha = [\zeta]$. These two vectors can be taken to correspond to the two orientations of the oriented sphere or plane represented by α, although we have not yet given a prescription as to how to make the correspondence. To do this, we need to introduce one more coordinate. First, embed $\mathbf{R}_1^{n+2}$ into $\mathbf{P}^{n+2}$ by the embedding $z \mapsto [(z, 1)]$. If $\zeta \in W^{n+1}$, then

$$-\zeta_1^2 + \zeta_2^2 + \cdots + \zeta_{n+2}^2 = 1,$$

so the point $[(\zeta, 1)]$ in $\mathbf{P}^{n+2}$ lies on the quadric Q^{n+1} in $\mathbf{P}^{n+2}$ given in homogeneous coordinates by the equation

$$\langle x, x \rangle = -x_1^2 + x_2^2 + \cdots + x_{n+2}^2 - x_{n+3}^2 = 0. \tag{2.11}$$

The manifold Q^{n+1} is called the *Lie quadric*, and the scalar product determined by the quadratic form in (2.11) is called the *Lie metric* or *Lie scalar product*. We will let $\{e_1, \ldots, e_{n+3}\}$ denote the standard orthonormal basis for the scalar product space $\mathbf{R}_2^{n+3}$ with metric $\langle , \rangle$. Here e_1 and e_{n+3} are timelike and the rest are spacelike.

We shall now see how points on Q^{n+1} correspond to the set of oriented hyperspheres, oriented hyperplanes and point spheres in $\mathbf{R}^n \cup \{\infty\}$. Suppose that x is any point on the quadric with homogeneous coordinate $x_{n+3} \neq 0$. Then x can be represented by a vector of the form $(\zeta, 1)$, where the Lorentz scalar product $(\zeta, \zeta) = 1$. Suppose first that $\zeta_1 + \zeta_2 \neq 0$. Then in Möbius geometry $[\zeta]$ represents a sphere in $\mathbf{R}^n$. If as in equation (2.6), we represent $[\zeta]$ by a vector of the form

$$\xi = \left(\frac{1 + p \cdot p - r^2}{2}, \frac{1 - p \cdot p + r^2}{2}, p \right),$$

then $(\xi, \xi) = r^2$. Thus ζ must be one of the vectors $\pm \xi / r$. In $\mathbf{P}^{n+2}$, we have

$$[(\zeta, 1)] = [(\pm \xi / r, 1)] = [(\xi, \pm r)].$$

We can interpret the last coordinate as a signed radius of the sphere with center p and unsigned radius $r > 0$. In order to be able to interpret this geometrically, we adopt the convention that a positive signed radius corresponds to the orientation of the sphere represented by the inward field of unit normals, and a negative signed radius corresponds to the orientation given by the outward field of unit normals. Hence, the two orientations of the sphere in $\mathbf{R}^n$ with center p and unsigned radius $r > 0$ are represented by the two projective points,

$$\left[\left(\frac{1 + p \cdot p - r^2}{2}, \frac{1 - p \cdot p + r^2}{2}, p, \pm r \right) \right] \tag{2.12}$$

in Q^{n+1}. Next if $\zeta_1 + \zeta_2 = 0$, then $[\zeta]$ represents a hyperplane in $\mathbf{R}^n$, as in equation (2.8). For $\zeta = (h, -h, N)$, with $|N| = 1$, we have $(\zeta, \zeta) = 1$. Then the two projective points on Q^{n+1} induced by ζ and $-\zeta$ are

$$[(h, -h, N, \pm 1)]. \tag{2.13}$$

These represent the two orientations of the plane with equation $u \cdot N = h$. We make the convention that $[(h, -h, N, 1)]$ corresponds to the orientation given by the field of unit normals N, while the orientation given by $-N$ corresponds to the point $[(h, -h, N, -1)] = [(-h, h, -N, 1)]$.

Thus far we have determined a bijective correspondence between the set of points x in Q^{n+1} with $x_{n+3} \neq 0$ and the set of all oriented spheres and planes in $\mathbf{R}^n$. Suppose now that $x_{n+3} = 0$, i.e., consider a point $[(z, 0)]$, for $z \in \mathbf{R}_1^{n+2}$. Then $\langle x, x \rangle = (z, z) = 0$, and $[z] \in \mathbf{P}^{n+1}$ is simply a point of the Möbius sphere Σ. Thus we have the following bijective correspondence between objects in Euclidean space and points on the Lie quadric:

Euclidean	**Lie**
points: $u \in \mathbf{R}^n$	$\left[\left(\frac{1+u \cdot u}{2}, \frac{1-u \cdot u}{2}, u, 0\right)\right]$
∞	$[(1, -1, 0, 0)]$
spheres: center p, signed radius r	$\left[\left(\frac{1+p \cdot p - r^2}{2}, \frac{1-p \cdot p + r^2}{2}, p, r\right)\right]$
planes: $u \cdot N = h$, unit normal N	$[(h, -h, N, 1)]$

(2.14)

In Lie sphere geometry, points are considered to be spheres of radius zero, or point spheres. From now on, we will use the term *Lie sphere* or simply "sphere" to denote an oriented sphere, oriented plane or a point sphere in $\mathbf{R}^n \cup \{\infty\}$. We will refer to the coordinates on the right side of equation (2.14) as the *Lie coordinates* of the corresponding point, sphere or plane. In the case of $\mathbf{R}^2$ and $\mathbf{R}^3$, respectively, these coordinates were classically called *pentaspherical* and *hexaspherical* coordinates (see [10]). At times it is useful to have formulas to convert Lie coordinates back into Cartesian equations for the corresponding Euclidean object. Suppose first that $[x]$ is a point on the Lie quadric with $x_1 + x_2 \neq 0$. Then $x = \rho y$, for some $\rho \neq 0$, where y is one of the standard forms on the right side of the table above. From the table, we see that $y_1 + y_2 = 1$, for all proper points and all spheres. Hence if we divide x by $x_1 + x_2$, the new vector will be in standard form, and we can read off the corresponding Euclidean object from the table. In particular, if $x_{n+3} = 0$, then $[x]$ represents the point $u = (u_3, \ldots, u_{n+2})$, where

$$u_i = x_i/(x_1 + x_2), \quad 3 \leq i \leq n + 2. \tag{2.15}$$

If $x_{n+3} \neq 0$, then $[x]$ represents the sphere with center $p = (p_3, \ldots, p_{n+2})$ and signed radius r given by

$$p_i = x_i/(x_1 + x_2), \quad 3 \leq i \leq n + 2; \quad r = x_{n+3}/(x_1 + x_2). \tag{2.16}$$

Finally, suppose that $x_1 + x_2 = 0$. If $x_{n+3} = 0$, then the equation $\langle x, x \rangle = 0$ forces x_i to be zero for $3 \leq i \leq n + 2$. Thus $[x] = [(1, -1, 0, \ldots, 0)]$, the improper point. If $x_{n+3} \neq 0$, we divide x by x_{n+3} to make the last coordinate 1. Then if we set $N = (N_3, \ldots, N_{n+2})$ and h according to

$$N_i = x_i/x_{n+3}, \quad 3 \leq i \leq n + 2; \quad h = x_1/x_{n+3}, \tag{2.17}$$

the conditions $\langle x, x \rangle = 0$ and $x_1 + x_2 = 0$ force N to have unit length. Thus $[x]$ corresponds to the hyperplane $u \cdot N = h$, with unit normal N and h as in equation (2.17).

2.4 Geometry of Hyperspheres in S^n and H^n

In some ways it is simpler to use the sphere S^n rather than $\mathbf{R}^n$ as the base space for the study of Möbius or Lie sphere geometry. This avoids the use of stereographic projection and the need to refer to an improper point or to distinguish between spheres and planes. Furthermore, the correspondence in the table in equation (2.14) can be reduced to a single formula (2.21) below.

As in Section 2.2, we consider S^n to be the unit sphere in $\mathbf{R}^{n+1}$, and then embed $\mathbf{R}^{n+1}$ into $\mathbf{P}^{n+1}$ by the canonical embedding ϕ. Then $\phi(S^n)$ is the Möbius sphere Σ, given by the equation $(z, z) = 0$ in homogeneous coordinates. First we find the Möbius equation for the unoriented hypersphere in S^n with center $p \in S^n$ and spherical radius $\rho, 0 < \rho < \pi$. This hypersphere is the intersection of S^n with the hyperplane in $\mathbf{R}^{n+1}$ given by the equation

$$p \cdot y = \cos \rho, \quad 0 < \rho < \pi. \tag{2.18}$$

Let $[z] = \phi(y) = [(1, y)]$. Then

$$p \cdot y = \frac{-(z, (0, p))}{(z, e_1)}.$$

Thus equation (2.18) can be rewritten as

$$(z, (\cos \rho, p)) = 0. \tag{2.19}$$

Therefore, a point $y \in S^n$ is on the hyperplane determined by equation (2.18) if and only if $\phi(y)$ lies on the polar hyperplane in $\mathbf{P}^{n+1}$ of the point

$$[\xi] = [(\cos \rho, p)]. \tag{2.20}$$

To obtain the two oriented spheres determined by equation (2.18) note that

$$(\xi, \xi) = -\cos^2 \rho + 1 = \sin^2 \rho.$$

Noting that $\sin \rho \neq 0$, we let $\zeta = \pm \xi / \sin \rho$. Then the point $[(\zeta, 1)]$ is on the quadric Q^{n+1}, and

$$[(\zeta, 1)] = [(\xi, \pm \sin \rho)] = [(\cos \rho, p, \pm \sin \rho)].$$

We can incorporate the sign of the last coordinate into the radius and thereby arrange that the oriented sphere S with signed radius $\rho \neq 0$, $-\pi < \rho < \pi$, and center p corresponds to a point in Q^{n+1} as follows:

$$S \longleftrightarrow [(\cos \rho, p, \sin \rho)]. \tag{2.21}$$

The formula still makes sense if the radius $\rho = 0$, in which case it yields the point sphere $[(1, p, 0)]$. This one formula (2.21) plays the role of all the formulas given in equation (2.14) in the preceding section for the Euclidean case.

As in the Euclidean case, the orientation of a sphere S in S^n is determined by a choice of unit normal field to S in S^n. Geometrically, we take the positive radius in (2.21) to correspond to the field of unit normals which are tangent vectors to geodesics from p to $-p$. Each oriented sphere can be considered in two ways, with center p and signed radius ρ, $-\pi < \rho < \pi$, or with center $-p$ and the appropriate signed radius $\rho \pm \pi$.

Given a point $[x]$ in the quadric Q^{n+1}, we now determine the corresponding hypersphere in S^n. Multiplying by -1, if necessary, we may assume that the first homogeneous coordinate x_1 of x satisfies $x_1 \geq 0$. If $x_1 > 0$, then we see from (2.21) that the center p and signed radius ρ, $-\pi/2 < \rho < \pi/2$, satisfy

$$\tan \rho = x_{n+3}/x_1, \qquad p = (x_2, \ldots, x_{n+2})/(x_1^2 + x_{n+3}^2)^{1/2}. \qquad (2.22)$$

If $x_1 = 0$, then $x_{n+3} \neq 0$, so we can divide by x_{n+3} to obtain a point of the form $(0, p, 1)$. This corresponds to the oriented hypersphere with center p and signed radius $\pi/2$, which is a great sphere in S^n.

To treat oriented hyperspheres in hyperbolic space H^n, we let $\mathbf{R}_1^{n+1}$ be the Lorentz subspace of $\mathbf{R}_1^{n+2}$ spanned by the orthonormal basis $\{e_1, e_3, \ldots, e_{n+2}\}$. Then H^n is the hypersurface

$$\{y \in \mathbf{R}_1^{n+1} \mid (y, y) = -1, \ y_1 \geq 1\},$$

on which the restriction of the Lorentz metric $(,)$ is a positive definite metric of constant sectional curvature -1 (see [95, Vol. II, pp. 268–271] for more detail). The distance between two points p and q in H^n is given by

$$d(p, q) = \cosh^{-1}(-(p, q)).$$

Thus the equation for the unoriented sphere in H^n with center p and radius ρ is

$$(p, y) = -\cosh \rho. \qquad (2.23)$$

As before with S^n, we first embed $\mathbf{R}_1^{n+1}$ into $\mathbf{P}^{n+1}$ by the map

$$\psi(y) = [y + e_2].$$

Let $p \in H^n$ and let $z = y + e_2$ for $y \in H^n$. Then we have

$$(p, y) = (z, p)/(z, e_2).$$

Hence equation (2.23) is equivalent to the condition that $[z] = [y + e_2]$ lies on the polar hyperplane in $\mathbf{P}^{n+1}$ to

$$[\xi] = [p + \cosh \rho \ e_2].$$

Following exactly the same procedure as in the spherical case, we find that the oriented hypersphere S in H^n with center p and signed radius ρ corresponds to a point in Q^{n+1} as follows:

$$S \longleftrightarrow [p + \cosh \rho \ e_2 + \sinh \rho \ e_{n+3}]. \qquad (2.24)$$

There is also a stereographic projection τ with pole $-e_1$ from H^n onto the unit disk D^n in $\mathbf{R}^n = \mathrm{Span}\{e_3, \ldots, e_{n+2}\}$ given by

$$\tau(y_1, y_3, \ldots, y_{n+2}) = (y_3, \ldots, y_{n+2})/(y_1 + 1). \qquad (2.25)$$

The metric g induced on D^n in order to make τ an isometry is the usual Poincaré metric.

In Section 3.5, we will see that from the point of view of Klein's Erlangen Program, all three of these geometries, Euclidean, spherical and hyperbolic, are subgeometries of Lie sphere geometry.

2.5 Oriented Contact and Parabolic Pencils of Spheres

In Möbius geometry, the principal geometric quantity is the angle. In Lie sphere geometry, the corresponding central concept is that of oriented contact of spheres. Two oriented spheres S_1 and S_2 in $\mathbf{R}^n$ are in *oriented contact* if they are tangent to each other and they have the same orientation at the point of contact. (See Figures 2.4 and 2.5 for the two possibilities.)

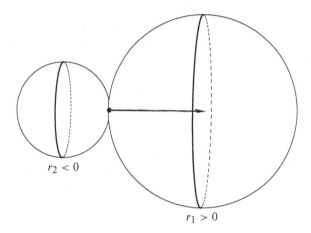

Fig. 2.4. Oriented contact of spheres, first case.

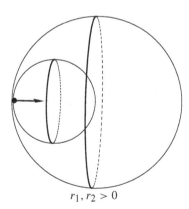

Fig. 2.5. Oriented contact of spheres, second case.

If p_1 and p_2 are the respective centers of S_1 and S_2, and r_1 and r_2 are their respective signed radii, then the analytic condition for oriented contact is

$$|p_1 - p_2| = |r_1 - r_2|. \tag{2.26}$$

An oriented sphere S with center p and signed radius r is in oriented contact with an oriented hyperplane π with unit normal N and equation $u \cdot N = h$ if π is tangent to S and their orientations agree at the point of contact. Analytically, this is just the equation

$$p \cdot N = r + h. \tag{2.27}$$

Two oriented planes π_1 and π_2 are in oriented contact if their unit normals N_1 and N_2 are the same. Two such planes can be thought of as two oriented spheres in oriented contact at the improper point.

A proper point u in $\mathbf{R}^n$ is in oriented contact sphere or plane if it lies on the sphere or plane. Finally, the improper point is in oriented contact with each plane, since it lies on each plane.

Suppose that S_1 and S_2 are two Lie spheres which are represented in the standard form given in equation (2.14) by $[k_1]$ and $[k_2]$. One can check directly that in all cases, the analytic condition for oriented contact is equivalent to the equation

$$\langle k_1, k_2 \rangle = 0. \tag{2.28}$$

Next we do some linear algebra to establish the important fact that the Lie quadric contains projective lines but no linear subspaces of higher dimension. We then show that the set of oriented spheres in $\mathbf{R}^n$ corresponding to the points on a line on Q^{n+1} forms a so-called *parabolic pencil* of spheres (see Figure 2.6). We also show that each parabolic pencil contains exactly one point sphere. Furthermore, if this point sphere is a proper point p in $\mathbf{R}^n$, then the pencil contains exactly one hyperplane π. The pencil consists of all oriented hyperspheres in oriented contact with π at the point p.

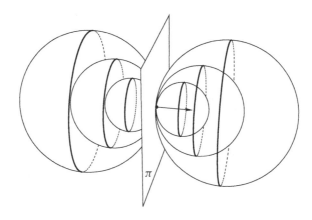

Fig. 2.6. Parabolic pencil of spheres.

The fundamental result needed from linear algebra is the following. Note that a subspace of a scalar product space is called *lightlike* if it consists of only lightlike vectors.

Theorem 2.3. *Let $(,)$ be a scalar product of signature $(n - k, k)$ on a real vector space V. Then the maximal dimension of a lightlike subspace of V is the minimum of the two numbers k and $n - k$.*

Proof. First, note that the theorem holds for scalar products having signature $(n-k, k)$ if and only if it holds for scalar products of signature $(k, n - k)$, since changing the signs of the quantities (e_i, e_i) for an orthonormal basis does not change the set of lightlike vectors.

Thus, we now assume that $k \leq n - k$ and do the proof by induction on the index k. The theorem is clearly true for scalar products of index 0, since the only lightlike vector is 0 itself. Assume now that the theorem holds for all spaces with a scalar product of index $k - 1$, and let V be a scalar product space of index $k \geq 1$. Let W be a lightlike subspace of V of maximal dimension, and let v be a timelike vector in V. Then the scalar product restricts to a scalar product of index $k - 1$ on the hyperplane $U = v^{\perp}$, and $W \cap U$ is a lightlike subspace of U. By the induction hypothesis, dim $W \cap U \leq k - 1$ and therefore, dim $W \leq k$, as desired. On the other hand, it is easy to exhibit a lightlike subspace of V of dimension k. Let $\{e_1, \ldots, e_n\}$ be an orthonormal basis for V with $\{e_1, \ldots, e_k\}$ timelike and the rest spacelike. For $1 \leq i \leq k$, let $v_i = e_i + e_{k+i}$. Then the span of $\{v_1, \ldots, v_k\}$ is a lightlike subspace of dimension k. $\qquad\square$

Corollary 2.4. *The Lie quadric contains projective lines but no linear subspaces of higher dimension.*

Proof. This follows immediately from Theorem 2.3, since a linear subspace of $\mathbf{P}^{n+2}$ of dimension $k - 1$ that lies on the quadric corresponds to a lightlike vector subspace of dimension k in $\mathbf{R}_2^{n+3}$. $\qquad\square$

Theorem 2.3 also implies the following result concerning the orthogonal complement of a line on the quadric. This was pointed out by Pinkall [146, p. 24].

Corollary 2.5. *Let ℓ be a line on $\mathbf{P}^{n+2}$ that lies on the quadric Q^{n+1}.*

(a) *If $[x] \in \ell^{\perp}$ and $[x]$ is lightlike, then $[x] \in \ell$.*
(b) *If $[x] \in \ell^{\perp}$ and $[x]$ is not on ℓ, then $[x]$ is spacelike.*

Proof.
(a) Suppose that $[x]$ is a lightlike point in $\ell^{\perp}$ but not on ℓ. Then the two-dimensional linear lightlike subspace spanned by $[x]$ and ℓ lies on the quadric, contradicting Corollary 2.4.

(b) Suppose that $[x]$ is in $\ell^{\perp}$ but not on ℓ. From (a) we know that $[x]$ is either spacelike or timelike. Suppose that $[x]$ is timelike. Then the Lie metric $\langle,\rangle$ has signature $(n + 1, 1)$ on the vector space $x^{\perp}$, and $x^{\perp}$ contains the two-dimensional lightlike vector space that projects to ℓ. This contradicts Theorem 2.3. $\qquad\square$

The next result establishes the relationship between the points on a line in Q^{n+1} and the corresponding parabolic pencil of spheres in $\mathbf{R}^n$.

Theorem 2.6.
(a) *The line in $\mathbf{P}^{n+2}$ determined by two points $[k_1]$ and $[k_2]$ of Q^{n+1} lies on Q^{n+1} if and only if the the spheres corresponding to $[k_1]$ and $[k_2]$ are in oriented contact, i.e., $\langle k_1, k_2 \rangle = 0$.*
(b) *If the line $[k_1, k_2]$ lies on Q^{n+1}, then the parabolic pencil of spheres in $\mathbf{R}^n$ corresponding to points on $[k_1, k_2]$ is precisely the set of all spheres in oriented contact with both $[k_1]$ and $[k_2]$.*

Proof.
  (a) The line $[k_1, k_2]$ consists of the points of the form $[\alpha k_1 + \beta k_2]$, where α and β are any two real numbers, at least one of which is not zero. Since $[k_1]$ and $[k_2]$ are on Q^{n+1}, we have

$$\langle \alpha k_1 + \beta k_2, \alpha k_1 + \beta k_2 \rangle = 2\alpha\beta \langle k_1, k_2 \rangle.$$

Thus the line is contained in the quadric if and only if $\langle k_1, k_2 \rangle = 0$.
  (b) Let $[\alpha k_1 + \beta k_2]$ be any point on the line. Since $\langle k_1, k_2 \rangle = 0$ by (a), we easily compute that $[\alpha k_1 + \beta k_2]$ is orthogonal to both $[k_1]$ and $[k_2]$. Hence, the corresponding sphere is in oriented contact with the spheres corresponding to $[k_1]$ and $[k_2]$. Conversely, suppose that the sphere corresponding to a point $[k]$ on the quadric is in oriented contact with the spheres corresponding to $[k_1]$ and $[k_2]$. Then $[k]$ is orthogonal to every point on the line $[k_1, k_2]$, and so $[k]$ is on the line $[k_1, k_2]$ by Corollary 2.5(a). $\square$

  As we have noted in the proofs of the previous results, given any timelike point $[z]$ in $\mathbf{P}^{n+2}$, the scalar product $\langle , \rangle$ has signature $(n+1, 1)$ on $z^\perp$. Hence, $z^\perp$ intersects Q^{n+1} in a Möbius space. We now show that any line on the quadric intersects such a Möbius space at exactly one point.

Corollary 2.7. *Let $[z]$ be a timelike point in $\mathbf{P}^{n+2}$ and ℓ a line that lies on Q^{n+1}. Then ℓ intersects $z^\perp$ at exactly one point.*

Proof. Any line in projective space intersects a hyperplane in at least one point. We simply must show that ℓ is not contained in $z^\perp$. But this follows from Theorem 2.3, since $\langle , \rangle$ has signature $(n + 1, 1)$ on $z^\perp$, and therefore $z^\perp$ cannot contain the two-dimensional lightlike vector space that projects to ℓ. $\square$

  As a consequence, we obtain the following corollary.

Corollary 2.8. *Every parabolic pencil contains exactly one point sphere. Furthermore, if the point sphere is a proper point, then the pencil contains exactly one plane.*

Proof. The point spheres are precisely the points of intersection of Q^{n+1} with $e_{n+3}^\perp$. Thus each parabolic pencil contains exactly one point sphere by Corollary 2.7. The hyperplanes correspond to the points in the intersection of Q^{n+1} with $(e_1 - e_2)^\perp$. The line ℓ on the quadric corresponding to the given parabolic pencil intersects this hyperplane at exactly one point unless ℓ is contained in the hyperplane. But ℓ is contained in $(e_1 - e_2)^\perp$ if and only if the improper point $[e_1 - e_2]$ is in $\ell^\perp$. By Corollary 2.5(a), this implies that the point $[e_1 - e_2]$ is on ℓ. Hence, if the point sphere of the pencil is not the improper point, then the pencil contains exactly one hyperplane.   □

By Corollary 2.8 and Theorem 2.6, we see that if the point sphere in a parabolic pencil is a proper point p in $\mathbf{R}^n$, then the pencil consists precisely of all spheres in oriented contact with a certain oriented plane π at p. Thus, one can identify the parabolic pencil with the point (p, N) in the unit tangent bundle to $\mathbf{R}^n$, where N is the unit normal to the oriented plane π. If the point sphere of the pencil is the improper point, then the pencil must consist entirely of planes. Since these planes are all in oriented contact, they all have the same unit normal N. Thus the pencil can be identified with the point (∞, N) in the unit tangent bundle to $\mathbf{R}^n \cup \{\infty\} = S^n$.

It is also useful to have this correspondence between parabolic pencils and elements of the unit tangent bundle $T_1 S^n$ expressed in terms of the spherical metric on S^n. Suppose that ℓ is a line on the quadric. From Corollary 2.7 and equation (2.21), we see that ℓ intersects both $e_1^\perp$ and $e_{n+3}^\perp$ at exactly one point. So the corresponding parabolic pencil contains exactly one point sphere and one great sphere, represented respectively by the points,

$$[k_1] = [(1, p, 0)], \quad [k_2] = [(0, \xi, 1)].$$

The fact that $\langle k_1, k_2 \rangle = 0$ is equivalent to the condition $p \cdot \xi = 0$, i.e., ξ is tangent to S^n at p. Hence the parabolic pencil of spheres corresponding to ℓ can be identified with the point (p, ξ) in $T_1 S^n$. The points on the line ℓ can be parametrized as

$$[K_t] = [\cos t \, k_1 + \sin t \, k_2] = [(\cos t, \cos t \, p + \sin t \, \xi, \sin t)].$$

From equation (2.21), we see that $[K_t]$ corresponds to the sphere in S^n with center

$$p_t = \cos t \, p + \sin t \, \xi, \tag{2.29}$$

and signed radius t. These are precisely the spheres through p in oriented contact with the great sphere corresponding to $[k_2]$. Their centers lie along the geodesic in S^n with initial point p and initial velocity vector ξ.

3

Lie Sphere Transformations

In this chapter, we study the sphere geometries of Lie, Möbius and Laguerre from the point of view of Klein's Erlangen Program. In each case, we determine the group of transformations which preserve the fundamental geometric properties of the space. All of these groups are quotient groups or subgroups of some orthogonal group determined by an indefinite scalar product on a real vector space. As a result, the theorem of Cartan and Dieudonné, proven in Section 3.2, implies that each of these groups is generated by inversions. In Section 3.3, we give a geometric description of Möbius inversions. This is followed by a treatment of Laguerre geometry in Section 3.4. Finally, in Section 3.5, we show that the Lie sphere group is generated by the union of the groups of Möbius and Laguerre. There we also describe the place of Euclidean, spherical and hyperbolic metric geometries within the context of these more general geometries.

3.1 The Fundamental Theorem

A *Lie sphere transformation* is a projective transformation of $\mathbf{P}^{n+2}$ which takes Q^{n+1} to itself. In terms of the geometry of $\mathbf{R}^n$, a Lie sphere transformation maps Lie spheres to Lie spheres. (Here the term "Lie sphere" includes oriented spheres, oriented planes and point spheres.) Furthermore, since it is projective, a Lie sphere transformation maps lines on Q^{n+1} to lines on Q^{n+1}. Thus, it preserves oriented contact of spheres in $\mathbf{R}^n$. We will first show that the group G of Lie sphere transformations is isomorphic to $O(n+1, 2)/\{\pm I\}$, where $O(n+1, 2)$ is the group of orthogonal transformations of $\mathbf{R}_2^{n+3}$. We will then give Pinkall's [147] proof of the so-called "fundamental theorem of Lie sphere geometry," which states that any line preserving diffeomorphism of Q^{n+1} is the restriction to Q^{n+1} of a projective transformation, that is, a transformation of the space of oriented spheres which preserves oriented contact must be a Lie sphere transformation.

Recall that a linear transformation $A \in GL(n + 1)$ induces a projective transformation $P(A)$ on $\mathbf{P}^n$ defined by $P(A)[x] = [Ax]$. The map P is a homomorphism of $GL(n + 1)$ onto the group $PGL(n)$ of projective transformations of $\mathbf{P}^n$. It is well known (see, for example, Samuel [166, p. 6]) that the kernel of P is the group of all nonzero scalar multiples of the identity transformation I.

The fact that the group G is isomorphic to $O(n+1,2)/\{\pm I\}$ follows immediately from the following theorem. Here we let $\langle,\rangle$ denote the scalar product on $\mathbf{R}_k^n$.

Theorem 3.1. *Let A be a nonsingular linear transformation on the indefinite scalar product space $\mathbf{R}_k^n$, $1 \leq k \leq n-1$, such that A takes lightlike vectors to lightlike vectors.*

(a) *Then there is a nonzero constant λ such that $\langle Av, Aw \rangle = \lambda \langle v, w \rangle$ for all v, w in $\mathbf{R}_k^n$.*

(b) *Furthermore, if $k \neq n - k$, then $\lambda > 0$.*

Proof.

(a) Since $k \geq 1$ and $n - k \geq 1$, there exist both timelike and spacelike vectors in $\mathbf{R}_k^n$. Suppose that v is a unit timelike vector and w is a unit spacelike vector such that $\langle v, w \rangle = 0$. Then $v + w$ and $v - w$ are both lightlike. By the hypothesis of the theorem, $A(v + w)$ and $A(v - w)$ are both lightlike. Thus, we have

$$0 = \langle A(v + w), A(v + w) \rangle = \langle Av, Av \rangle + 2\langle Av, Aw \rangle + \langle Aw, Aw \rangle, \qquad (3.1)$$
$$0 = \langle A(v - w), A(v - w) \rangle = \langle Av, Av \rangle - 2\langle Av, Aw \rangle + \langle Aw, Aw \rangle.$$

If we subtract the second equation from the first, we get $\langle Av, Aw \rangle = 0$. Substitution of this into either of the equations above yields

$$-\langle Av, Av \rangle = \langle Aw, Aw \rangle = \lambda, \qquad (3.2)$$

for some real number λ. Now suppose that $\{v_1, \ldots, v_k, w_1, \ldots, w_{n-k}\}$ is an orthonormal basis for $\mathbf{R}_k^n$ with $v_1, \ldots, v_k$ timelike and $w_1, \ldots, w_{n-k}$ spacelike. We have already shown that $\langle Av_i, Aw_j \rangle = 0$ for all i and j. From equation (3.2) we get that $-\langle Av, Av \rangle = \langle Aw, Aw \rangle = \lambda$, for all i, j, since we can first hold v constant in (3.2) and vary w, then hold w constant and vary v. It remains to be shown that $\langle Av_i, Av_j \rangle = 0$ and $\langle Aw_i, Aw_j \rangle = 0$ for $i \neq j$. Consider the vector $(w_i + w_j)/\sqrt{2}$. Then w is a unit spacelike vector orthogonal to v_1. Hence, we have $\langle Aw, Aw \rangle = \lambda$, i.e.,

$$2\lambda = \langle A(w_i + w_j), A(w_i + w_j) \rangle = \langle Aw_i, Aw_i \rangle + 2\langle Aw_i, Aw_j \rangle + \langle Aw_j, Aw_j \rangle. \qquad (3.3)$$

Since $\langle Aw_i, Aw_i \rangle = \langle Aw_j, Aw_j \rangle = \lambda$, we have $\langle Aw_i, Aw_j \rangle = 0$ for $i \neq j$. A similar proof shows that $\langle Av_i, Av_j \rangle = 0$ for $i \neq j$. Therefore, the equation $\langle Ax, Ay \rangle = \lambda \langle x, y \rangle$ holds on an orthonormal basis, so it holds for all vectors.

To prove (b), note that $\langle,\rangle$ has signature $(k, n - k)$; so if $k \neq n - k$, then the Av_i must be timelike and the Aw_i spacelike, i.e., $\lambda > 0$. $\qquad \square$

Remark 3.2. In the case $k = n - k$, conclusion (b) does not necessarily hold. For example, the linear map T defined by $Tv_i = w_i$, $Tw_i = v_i$, for $1 \leq i \leq k$, preserves lightlike vectors, but the corresponding $\lambda = -1$.

From Theorem 3.1 we immediately obtain the following corollary.

Corollary 3.3.
(a) *The group G of Lie sphere transformations is isomorphic to $O(n+1, 2)/\{\pm I\}$.*
(b) *The group H of Möbius transformations is isomorphic to $O(n+1, 1)/\{\pm I\}$.*

Proof.

(a) Suppose $\alpha = P(A)$ is a Lie sphere transformation. By Theorem 3.1, we have $\langle Av, Aw \rangle = \lambda \langle v, w \rangle$ for all v, w in $\mathbf{R}_2^{n+3}$, where λ is a positive constant. Set B equal to $A/\sqrt{\lambda}$. Then B is in $O(n+1, 2)$ and $\alpha = P(B)$. Thus, every Lie sphere transformation can be represented by an orthogonal transformation. Conversely, if $B \in O(n+1, 2)$, then $P(B)$ is clearly a Lie sphere transformation. Now let $\Psi : O(n+1, 2) \to G$ be the restriction of the homomorphism P to $O(n+1, 2)$. Then Ψ is surjective with kernel equal to the intersection of the kernel of P with $O(n+1, 2)$, i.e., kernel $\Psi = \{\pm I\}$.

(b) This follows from Theorem 3.1 in the same manner as (a) with the Lorentz metric being used instead of the Lie metric. □

Remark 3.4 (on Möbius transformations in Lie sphere geometry). A Möbius transformation is a transformation on the space of unoriented spheres, i.e., the space of projective classes of spacelike vectors in $\mathbf{R}_1^{n+2}$. Hence, each Möbius transformation naturally induces two Lie sphere transformations on the space Q^{n+1} of oriented spheres. Specifically, if A is in $O(n+1, 1)$, then we can extend A to a transformation B in $O(n+1, 2)$ by setting $B = A$ on $\mathbf{R}_1^{n+2}$ and $B(e_{n+3}) = e_{n+3}$. In terms of matrix representation with respect to the standard orthonormal basis, B has the form

$$B = \begin{bmatrix} A & 0 \\ 0 & 1 \end{bmatrix}. \tag{3.4}$$

Note that while A and $-A$ induce the same Möbius transformation, the Lie transformation $P(B)$ is not the same as the Lie transformation $P(C)$ induced by the matrix

$$C = \begin{bmatrix} -A & 0 \\ 0 & 1 \end{bmatrix} \simeq \begin{bmatrix} A & 0 \\ 0 & -1 \end{bmatrix},$$

where $\simeq$ denotes equivalence as projective transformations. Hence, the Möbius transformation $P(A) = P(-A)$ induces two Lie transformations, $P(B)$ and $P(C)$. Finally, note that $P(B) = \Gamma P(C)$, where Γ is the Lie transformation represented in matrix form by

$$\Gamma = \begin{bmatrix} I & 0 \\ 0 & -1 \end{bmatrix} \simeq \begin{bmatrix} -I & 0 \\ 0 & 1 \end{bmatrix}.$$

From equation (2.14) of Chapter 2, p. 16, we see that Γ has the effect of changing the orientation of every oriented sphere or plane. We will call Γ the *change of orientation transformation*, although the German "Richtungswechsel" is certainly more economical. Hence, the two Lie sphere transformations induced by the Möbius transformation $P(A)$ differ by this change of orientation factor. Thus, the group of Lie transformations induced from Möbius transformations is isomorphic to $O(n+1, 1)$ and is a double covering of the Möbius group H. This group consists of those Lie transformations that map $[e_{n+3}]$ to itself. Since such a transformation must also take $e_{n+3}^{\perp}$ to itself, this is precisely the group of Lie transformations which take point spheres to point spheres. When working in the context of Lie sphere geometry, we will often refer to these transformations as "Möbius transformations."

We now turn to the proof of the fundamental theorem of Lie sphere geometry. Lie [104, p. 186] proved this result in his thesis for $n = 2$. (See also Lie–Scheffers [105, p. 437] or Blaschke [10, p. 211].) The following proof is due to Pinkall [147, p. 431].

Theorem 3.5. *Every line preserving diffeomorphism of Q^{n+1} is the restriction to Q^{n+1} of a Lie sphere transformation.*

Proof. The key observation in this proof is that a line preserving diffeomorphism of Q^{n+1} corresponds to a conformal transformation of Q^{n+1} endowed with its natural pseudo-Riemannian metric of signature $(n, 1)$. Theorem 3.5 then follows from the fact that such a conformal transformation must be the restriction to Q^{n+1} of a projective transformation of $\mathbf{P}^{n+2}$ induced by an orthogonal transformation of $\mathbf{R}_2^{n+3}$. This is a generalization of the fact that a conformal transformation of S^n must be a Möbius transformation, since S^n is just the quadric obtained by projection of the light cone in $\mathbf{R}_1^{n+2}$ (see Section 2.2). In fact, given any scalar product space $\mathbf{R}_{k+1}^{m+2}$, the projective quadric Q_k^m obtained by projecting the light cone in $\mathbf{R}_{k+1}^{m+2}$ has a natural pseudo-Riemannian metric of signature $(m - k, k)$. With this metric Q_k^m is a conformally flat pseudo-Riemannian symmetric space, called the *standard projective quadric* of signature $(m - k, k)$. (In this notation, our Q^{n+1} is Q_1^{n+1} and S^n is Q_0^n.) Cahen and Kerbrat [14, pp. 327–331] give a proof that the group of conformal transformations of Q_k^m is isomorphic to $O(m - k + 1, k + 1)/\{\pm I\}$ which works for all signatures at once. Here we will construct the standard metric for our quadric Q^{n+1} and demonstrate that a line preserving diffeomorphism of Q^{n+1} determines a conformal transformation of Q^{n+1} with respect to this metric. Let V^{n+2} be the light cone in $\mathbf{R}_2^{n+3}$, and let

$$M^{n+1} = \{x \in \mathbf{R}_2^{n+3} \mid x_1^2 + x_{n+3}^2 = 1, \ x_2^2 + \cdots + x_{n+2}^2 = 1\}.$$

The manifold M^{n+1} is the intersection of V^{n+2} with the hypersphere of radius $\sqrt{2}$ in the Euclidean metric on $\mathbf{R}^{n+3}$. It is clearly diffeomorphic to $S^1 \times S^n$, where S^1 is the circle with equation $x_1^2 + x_{n+3}^2 = 1$ in the timelike plane spanned by e_1 and e_{n+3}, and S^n is the n-sphere given by the equation

$$x_2^2 + \cdots + x_{n+2}^2 = 1,$$

in the Euclidean space $\mathbf{R}^{n+1}$ spanned by $\{e_2, \ldots, e_{n+2}\}$. The manifold M^{n+1} is a double covering of Q^{n+1}, and the fiber containing the point $x \in M^{n+1}$ is the orbit of x under the action of the group $\mathbf{Z}_2 = \{\pm I\}$.

Suppose that $x = (w, z)$ is an arbitrary point of $S^1 \times S^n = M^{n+1}$. Choose an orthornormal basis $\{u_1, \ldots, u_{n+3}\}$ of $\mathbf{R}_2^{n+3}$ with u_1 and u_{n+3} timelike and the rest spacelike such that $u_1 = w$ and $u_2 = z$. Then the tangent space

$$T_x M^{n+1} = T_w S^1 \times T_z S^n = \text{Span}\{u_3, \ldots, u_{n+3}\}.$$

Thus the restriction h of $\langle,\rangle$ to $T_x M^{n+1}$ has signature $(n, 1)$. The metric h is invariant under the action of $\mathbf{Z}_2$, so it induces a pseudo-Riemannian metric g of signature $(n, 1)$ on Q^{n+1}. Let π be the projection $x \mapsto [x]$ from $\mathbf{R}_2^{n+3}$ to $\mathbf{P}^{n+2}$. Then $\pi^* g$ determines a tensor field on the punctured cone $V^{n+2} - \{0\}$ which is invariant under central dilatations $x \mapsto ax, a \neq 0$, and coincides with h on M^{n+1}. This metric is

$$\pi^* g(Y, Z) = 2\langle Y, Z \rangle / |x|^2, \tag{3.5}$$

where $|x|$ is the Euclidean length of x in $\mathbf{R}_2^{n+3}$, and Y, Z are tangent to V^{n+2} at x. Thus, one can also consider g to be induced from the metric $\pi^* g$ on the punctured cone.

The metric g can be shown to be conformally flat as follows. Let

$$\{u_1, \ldots, u_{n+3}\}$$

be any orthornormal basis of $\mathbf{R}_2^{n+3}$ with u_1 and u_{n+3} timelike. Let U be the open subset of points $[x]$ in Q^{n+1} whose homogeneous coordinates with respect to this basis satisfy $x_1 + x_2 \neq 0$. We will now show that U is conformally diffeomorphic to the Lorentz space $\mathbf{R}_1^{n+1}$ spanned by $\{u_3, \ldots, u_{n+3}\}$. By taking an appropriate scalar multiple, we may assume that the homogeneous coordinates

$$x = (x_1, \ldots, x_{n+3})$$

of a point $[x]$ in U satisfy $x_1 + x_2 = 1$. Let $X = (x_3, \ldots, x_{n+3})$ and let

$$(X, X) = x_3^2 + \cdots + x_{n+2}^2 - x_{n+3}^2 \tag{3.6}$$

be the restriction of $\langle,\rangle$ to the Lorentz space $\mathbf{R}_1^{n+1}$. Then,

$$0 = \langle x, x \rangle = -x_1^2 + x_2^2 + (X, X) = -x_1 + x_2 + (X, X),$$

since $x_1 + x_2 = 1$. Hence, we have $x_1 - x_2 = (X, X)$, and we can solve for x_1 and x_2 as

$$x_1 = (1 + (X, X))/2, \quad x_2 = (1 - (X, X))/2. \tag{3.7}$$

Thus, we have a diffeomorphism $\beta : \mathbf{R}_1^{n+1} \to U$ defined by $\beta(X) = [\psi(X)]$, where $\psi(X) = (x_1, x_2, X)$ for x_1 and x_2 as in (3.7).

To show that β is conformal, we consider the map $\psi : \mathbf{R}_1^{n+1} \to V^{n+2}$ and use the metric $\pi^* g$ given by equation (3.5). Let Y be a tangent vector to $\mathbf{R}_1^{n+1}$ at the point X. From equation (3.7), we compute the differential $d\psi$ to be

$$d\psi(Y) = ((X, Y), -(X, Y), Y).$$

If Z is another tangent vector to $\mathbf{R}_1^{n+1}$ at X, then

$$\langle d\psi(Y), d\psi(Z) \rangle = -(X, Y)(X, Z) + (X, Y)(X, Z) + (Y, Z) = (Y, Z).$$

By equation (3.5) and the equation above, we have

$$\pi^* g(d\psi(Y), d\psi(Z)) = 2\langle Y, Z\rangle / |\psi(X)|^2,$$

and β is conformal.

Now we want to show that the lines in U that lie on the quadric are precisely the images under β of lightlike lines in $\mathbf{R}_1^{n+1}$. Consider two points in U with homogeneous coordinates $x = (x_1, x_2, X)$ and $y = (y_1, y_2, Y)$ satisfying the equation

$$x_1 + x_2 = y_1 + y_2 = 1.$$

The line $[x, y]$ lies on the quadric precisely when $\langle x, y\rangle = 0$. A direct computation using equation (3.7) shows that

$$\langle x, y\rangle = -\langle X - Y, X - Y\rangle / 2.$$

Hence, the line $[x, y]$ lies on the quadric Q^{n+1} if and only if $X - Y$ is lightlike, i.e., the line $[X, Y]$ is a lightlike line in $\mathbf{R}_1^{n+1}$.

Since the diffeomorphism β is conformal, the paragraph above implies that lightlike vectors in the tangent space $T_q Q^{n+1}$ at any point $q \in U$ are precisely the tangent vectors to lines through q that lie on the quadric. The same can be said for all points of Q^{n+1}, since every point of the the quadric lies in an open set similar to U, for an appropriate choice of homogeneous coordinate basis $\{u_1, \ldots, u_{n+3}\}$.

We now complete the proof of Theorem 3.5. Let ϕ be a line preserving diffeomorphism of Q^{n+1}. Then its differential $d\phi$ takes lightlike vectors in the tangent space $T_q Q^{n+1}$ to lightlike vectors in the tangent space of Q^{n+1} at $\phi(q)$. Each of these spaces is isomorphic to $\mathbf{R}_1^{n+1}$. By applying Theorem 3.1 to the linear map $d\phi$, we conclude that ϕ is conformal. Then by the classification of conformal transformal transformations of (Q^{n+1}, g) in Cahen–Kerbrat [14, pp. 327–331], ϕ is the restriction to Q^{n+1} of a projective transformation of $\mathbf{P}^{n+2}$ taking Q^{n+1} to itself, i.e., a Lie sphere transformation. $\square$

3.2 Generation of the Lie Sphere Group by Inversions

In this section, we will show that the group G of Lie sphere transformations and the group H of Möbius transformations are generated by inversions. This follows from the fact that the corresponding orthogonal groups are generated by reflections in hyperplanes. In fact, every orthogonal transformation on $\mathbf{R}_k^n$ is a product of at most n reflections, a result due to Cartan and Dieudonné. Our treatment of this result follows from E. Artin's book [4, Chapter 3]. (See also Cartan [20, pp. 10–12].)

For the moment, let $\langle,\rangle$ denote the scalar product of signature $(n - k, k)$ on $\mathbf{R}_k^n$. A hyperplane π in $\mathbf{R}_k^n$ is called *nondegenerate* if the scalar product restricts to a nondegenerate form on π. From Theorem 2.2 of Chapter 2, p. 10, we know that a hyperplane π is nondegenerate if and only if its pole ξ is not lightlike. Now let ξ be a unit spacelike or unit timelike vector in $\mathbf{R}_k^n$. The *reflection* Ω_π of $\mathbf{R}_k^n$ in the hyperplane π with pole ξ is defined by the formula

$$\Omega_\pi x = x - \frac{2\langle x, \xi \rangle \xi}{\langle \xi, \xi \rangle}. \tag{3.8}$$

Note that we do not define reflection in degenerate hyperplanes, i.e., those which have lightlike poles. It is clear that Ω_π fixes every point in π and that $\Omega_\pi \xi = -\xi$. A direct computation shows that Ω_π is in $O(n - k, k)$ and that $\Omega_\pi^2 = I$.

In the proof of the theorem of Cartan and Dieudonné, we need Lemma 3.6 below concerning the special case of $\mathbf{R}_k^{2k}$, where the metric has signature (k, k). In that case, let $\{e_1, \ldots, e_{2k}\}$ be an orthonormal basis with $e_1, \ldots, e_k$ spacelike and $e_{k+1}, \ldots, e_{2k}$ timelike. One can naturally choose a basis

$$\{v_1, \ldots, v_k, w_1, \ldots, w_k\}$$

of lightlike vectors given by

$$v_i = (e_i + e_{k+i})/\sqrt{2}, \quad w_i = (e_i - e_{k+i})/\sqrt{2}. \tag{3.9}$$

Note that the scalar products of these vectors satisfy

$$\langle v_i, v_j \rangle = 0, \quad \langle w_i, w_j \rangle = 0, \quad \langle v_i, w_j \rangle = \delta_{ij}, \tag{3.10}$$

for all i, j.

Let V be the lightlike subspace of dimension k spanned by $v_1, \ldots, v_k$. Suppose that U is any other lightlike subspace of dimension k in $\mathbf{R}_k^{2k}$. Let α be any linear isomorphism of V onto U. Since both spaces are lightlike, α is trivially an isometry. By Witt's theorem (see Artin [4, p. 121]), there is an orthogonal transformation ϕ of $\mathbf{R}_k^{2k}$ which extends α. The vectors $\phi(v_i)$ and $\phi(w_i)$ satisfy the same scalar product relations (3.10) as the v_i and w_i, and U is spanned by $\phi(v_1), \ldots, \phi(v_k)$. Using this, we can now prove the lemma.

Lemma 3.6. *Suppose that an orthogonal transformation σ fixes every vector in a lightlike subspace U of dimension k in $\mathbf{R}_k^{2k}$. Then σ has determinant one.*

Proof. As we noted above, there exists a basis $\{v_1, \ldots, v_k, w_1, \ldots, w_k\}$ of lightlike vectors in $\mathbf{R}_k^{2k}$ satisfying equation (3.10) such that U is the subspace spanned by $v_1, \ldots, v_k$. We now determine the matrix of σ with respect to this basis. We are given that $\sigma v_i = v_i$ for $1 \leq i \leq k$. Let

$$\sigma w_j = \sum_{h=1}^{k} a_{hj} v_h + \sum_{m=1}^{k} b_{mj} w_m. \tag{3.11}$$

Since σ preserves scalar products, we have from equations (3.10) and (3.11),

$$b_{ij} = \langle v_i, \sigma w_j \rangle = \langle \sigma v_i, \sigma w_j \rangle = \langle v_i, w_j \rangle = \delta_{ij}.$$

Since the matrix for σ with respect to the basis $\{v_1, \ldots, v_k, w_1, \ldots, w_k\}$ has zeroes below the diagonal and ones along the diagonal, σ has determinant one. $\square$

We now prove the theorem of Cartan and Dieudonné.

Theorem 3.7. *Every orthogonal transformation of $\mathbf{R}_k^n$ is the product of at most n reflections in hyperplanes.*

Proof. The proof is by induction on the dimension n. In the case $n = 1$, the only orthogonal transformations are $\pm I$. The identity is the product of zero reflections and $-I$ is a reflection if $n = 1$. We now assume that any orthogonal transformation on a scalar product space of dimension $n - 1$ is the product of at most $n - 1$ reflections. Let σ be a given orthogonal transformation on $\mathbf{R}_k^n$. We must show that σ is the product of at most n reflections. We need to distinguish four cases:

CASE 1. *There exists a nonlightlike vector v which is fixed by σ.*

In this case, let $\pi = v^{\perp}$. Since σ is orthogonal and fixes v, we have $\sigma\pi = \pi$. Let λ be the restriction of σ to π. By the induction hypothesis, λ is the product of at most $n - 1$ reflections in hyperplanes of the space π, say $\lambda = \Omega_1 \cdots \Omega_r$, where $r \leq n - 1$. Each Ω_i extends to a reflection in a hyperplane of $\mathbf{R}_k^n$ by setting $\Omega_i(v) = v$. Then, since σ and the product $\Omega_1 \cdots \Omega_r$ agree on π and on v, they are equal. Thus, in this case, σ is the product of at most $n - 1$ reflections.

CASE 2. *There is a nonlightlike vector v such that $\sigma v - v$ is nonlightlike.*

Let π be the hyperplane $(\sigma v - v)^{\perp}$. Since σ is orthogonal, we have

$$\langle \sigma v + v, \sigma v - v \rangle = \langle \sigma v, \sigma v \rangle - \langle v, v \rangle = 0.$$

Thus, $\sigma v + v$ is in π, and we have

$$\Omega_{\pi}(\sigma v + v) = \sigma v + v, \quad \Omega_{\pi}(\sigma v - v) = v - \sigma v, \tag{3.12}$$

where Ω_{π} is the reflection in the hyperplane π. Adding the two equations in (3.12) and using the linearity of Ω_{π}, we get $\Omega_{\pi} \sigma v = v$, for the nonlightlike vector v. Now by Case 1, we know that $\Omega_{\pi} \sigma$ is the product of at most $n - 1$ reflections $\Omega_1 \cdots \Omega_r$. Thus,

$$\sigma = \Omega_{\pi} \Omega_1 \cdots \Omega_r$$

is the product of at most n reflections.

CASE 3. $n = 2$.

By Cases 1 and 2, we need only handle the case where the signature of the metric is $(1, 1)$. In that case, let $\{u, v\}$ be a basis of lightlike vectors as in equation (3.9) satisfying

$$\langle u, u \rangle = 0, \quad \langle w, w \rangle = 0, \quad \langle u, w \rangle = 1. \tag{3.13}$$

Since σu is lightlike, it must be a multiple of u or w. We handle the two possibilities separately.

(a) Suppose that $\sigma u = aw$, for $a \neq 0$. Since σ is orthogonal, equation (3.13) implies that $\sigma w = a^{-1}u$. Then, $\sigma(u + aw) = aw + u$, and $u + aw$ is a fixed nonlightlike vector. So Case 1 applies.

(b) Suppose that $\sigma u = au$, for $a \neq 0$. Then equation (3.13) implies that $\sigma w = a^{-1}w$. If the number $a = 1$, then σ is the identity, and we are done, If $a \neq 1$, we consider $v = u + w$. Then v is nonlightlike, and

$$\sigma v - v = (a - 1)u + (a^{-1} - 1)w,$$

which is a nonlightlike vector. Hence, Case 2 applies. Thus the only remaining case to be handled is the following.

CASE 4. $n \geq 3$; *no nonlightlike vector is fixed by σ, and $\sigma v - v$ is lightlike for every nonlightlike vector v.*

Let u be any lightlike vector. We first show that $\sigma u - u$ must be lightlike. By Theorem 2.2 of Chapter 2, p. 10, we know that $\dim u^{\perp} = n - 1$. Since $n \geq 3$, we know that $n - 1$ is greater than $n/2$. Since the maximum possible dimension of a lightlike subspace is less than or equal to $n/2$ by Theorem 2.3 of Chapter 2, p. 21, we know that $u^{\perp}$ contains a nonlightlike vector v. Then since $\langle v, v \rangle \neq 0$, we have

$$\langle v + \varepsilon u, v + \varepsilon u \rangle = \langle v, v \rangle \neq 0,$$

for any real number ε. By our assumption, $\sigma v - v$ is lightlike and so also is

$$w = \sigma(v + \varepsilon u) - (v + \varepsilon u) = \sigma v - v + \varepsilon(\sigma u - u),$$

for every ε. Thus,

$$\langle w, w \rangle = 2\varepsilon \langle \sigma v - v, \sigma u - u \rangle + \varepsilon^2 \langle \sigma u - u, \sigma u - u \rangle = 0, \qquad (3.14)$$

for all ε. If we take $\varepsilon = 1$ and $\varepsilon = -1$ in equation (3.14) and add the equations, we get

$$2\langle \sigma u - u, \sigma u - u \rangle = 0,$$

so that $\sigma u - u$ is lightlike for any lightlike vector u.

Thus, we now have that $\sigma x - x$ is lightlike for every vector x in $\mathbf{R}_k^n$. Let W be the image of the linear transformation $\sigma - I$. Then W is a lightlike subspace of $\mathbf{R}_k^n$, and so the scalar product of any two vectors in W is zero. Now let $x \in \mathbf{R}_k^n$ and $y \in W^{\perp}$. Then $\sigma x - x$ and $\sigma y - y$ are in W, so

$$0 = \langle \sigma x - x, \sigma y - y \rangle = \langle \sigma x, \sigma y \rangle - \langle x, \sigma y \rangle - \langle \sigma x - x, y \rangle. \qquad (3.15)$$

Since $\sigma x - x$ is in W and y is in $W^{\perp}$, the last term is zero. Furthermore, since $\langle \sigma x, \sigma y \rangle$ equals $\langle x, y \rangle$, equation (3.15) reduces to

$$\langle x, y - \sigma y \rangle = 0.$$

Since this holds for all x in the scalar product space $\mathbf{R}_k^n$, we conclude that the vector $y - \sigma y = 0$, i.e., $\sigma y = y$ for all $y \in W^{\perp}$. Since no nonlightlike vectors are fixed by σ, this implies that $W^{\perp}$ consists entirely of lightlike vectors. Now we have two lightlike subspaces W and $W^{\perp}$. We know that the sum of their dimensions is n and that each has dimension at most $n/2$ by Theorems 2.2 (p. 10) and 2.3 (p. 21) of Chapter 2. Thus each space has dimension $n/2$ and the signature of the metric is (k, k) where $k = n/2$. Furthermore, since σ fixes every vector in $W^\perp$, Lemma 3.6 implies that the determinant of σ is equal to 1.

Hence, the theorem holds for all orthogonal transformations of $\mathbf{R}_k^n$, with the possible exception of transformations with determinant 1 on $\mathbf{R}_k^{2k}$. In particular, any orthogonal transformation of $\mathbf{R}_k^{2k}$ with determinant -1 is the product of at most $2k$ reflections. But the product of $2k$ reflections has determinant 1, so an orthogonal transformation of $\mathbf{R}_k^{2k}$ with determinant -1 is a product of less than $2k$ reflections. Now, let Ω be any reflection in a hyperplane in $\mathbf{R}_k^{2k}$. Then $\Omega\sigma$ has determinant -1, so it is the product $\Omega_1 \cdots \Omega_r$ of less than $2k$ reflections. Therefore,

$$\sigma = \Omega\Omega_1 \cdots \Omega_r$$

is the product of at most $2k$ reflections, as desired. $\square$

We now return to the context of Lie sphere geometry. The Lie sphere transformation induced by a reflection Ω_π in $O(n + 1, 2)$ is called a *Lie inversion*. Similarly, a Möbius transformation induced by a reflection in $O(n + 1, 1)$ is called a *Möbius inversion*. An immediate consequence of Corollary 3.3 and Theorem 3.7 is the following.

Theorem 3.8. *The Lie sphere group G and the Möbius group H are both generated by inversions.*

In the next two sections, we will give a geometric description of these inversions and other important types of Lie sphere transformations.

3.3 Geometric Description of Inversions

In this section, we begin with a geometric description of Möbius inversions. This is followed by a more general discussion of Lie sphere transformations, which leads naturally to the sphere geometry of Laguerre treated in the next section.

An orthogonal transformation in $O(n + 1, 1)$ induces a projective transformation on $\mathbf{P}^{n+1}$ which maps the Möbius sphere Σ to itself. A Möbius inversion is the projective transformation induced by a reflection Ω_π in $O(n+1, 1)$. For the sake of brevity, we will also denote this projective transformation by Ω_π instead of $P(\Omega_\pi)$. Let ξ be a spacelike point in $\mathbf{P}^{n+1}$ with polar hyperplane π. The hyperplane π intersects the Möbius sphere Σ in a hypersphere S^{n-1}. The Möbius inversion Ω_π, when interpreted as a transformation on $\mathbf{R}^n$, is just ordinary inversion in the hypersphere S^{n-1}. We will now recall the details of this transformation.

Since the Möbius sphere is homogeneous, all inversions in planes with spacelike poles act in essentially the same way. Let us consider the special case where S^{n-1} is the sphere of radius $r > 0$ centered at the origin in $\mathbf{R}^n$. Then by formula

(2.6) of Chapter 2, p. 12, the spacelike point ξ in $\mathbf{P}^{n+1}$ corresponding to S^{n-1} has homogeneous coordinates

$$\xi = (1 - r^2, 1 + r^2, 0)/2.$$

Let u be a point in $\mathbf{R}^n$ other than the origin. By equation (2.3) of Chapter 2, p. 11, the point u corresponds to the point in $\mathbf{P}^{n+1}$ with homogeneous coordinates

$$x = (1 + u \cdot u, 1 - u \cdot u, 2u)/2.$$

The formula for Ω_π in homogeneous coordinates is

$$\Omega_\pi x = x - \frac{2(x, \xi)}{(\xi, \xi)} \xi, \tag{3.16}$$

where $(,)$ is the Lorentz metric. A straightforward calculation shows that $\Omega_\pi x$ is the point in $\mathbf{P}^{n+1}$ with homogeneous coordinates

$$(1 + v \cdot v, 1 - v \cdot v, 2v)/2,$$

where $v = (r^2/|u|^2)u$. Thus, the Euclidean transformation induced by Ω_π maps u to the point v on the ray through u from the origin satisfying the equation $|u||v| = r^2$. From this, it is clear that the fixed points of Ω_π are precisely the points of the sphere S^{n-1}. Viewed in the projective context, this is immediately clear from equation (3.16), since $\Omega_\pi x = x$ if and only if $(x, \xi) = 0$. In general, an inversion of $\mathbf{R}^n$ in the hypersphere of radius r centered at a point p maps a point $u \neq p$ to the point v on the ray through u from p satisfying

$$|u - p||v - p| = r^2.$$

Another special case is when the unit spacelike vector ξ lies in the Euclidean space $\mathbf{R}^n$ spanned by $\{e_3, \ldots, e_{n+2}\}$. Then the "sphere" corresponding to $[\xi]$ according to formula (2.8) of Chapter 2, p. 13, is the hyperplane V through the origin in $\mathbf{R}^n$ perpendicular to ξ. In this case the Möbius inversion Ω_π is just ordinary Euclidean reflection in the hyperplane V.

It is also instructive to study inversion as a map from the Möbius sphere Σ to itself. Suppose that ξ is a spacelike point in $\mathbf{P}^{n+1}$ and x is a point on Σ which is not on the polar hyperplane π of ξ. The line $[x, \xi]$ in $\mathbf{P}^{n+1}$ intersects Σ in precisely two points, x and $\Omega_\pi x$ (see Figure 3.1), and Ω_π simply exchanges these two points.

Given a spacelike point η, the sphere polar to η is taken by Ω_π to the sphere polar to $\Omega_\pi \eta$, since Ω_π is an orthogonal transformation. Thus the sphere polar to η is taken to itself by Ω_π if and only if η is fixed by Ω_π, i.e., $(\xi, \eta) = 0$. Geometrically this means that the sphere polar to η is orthogonal to the sphere of inversion.

If ξ is a timelike point in $\mathbf{P}^{n+1}$, then formula (3.16) still makes sense, although the polar hyperplane π to ξ does not intersect Σ, and thus $(x, \xi) \neq 0$ for each $x \in \Sigma$. Given a point $x \in \Sigma$, the line $[x, \xi]$ intersects Σ in precisely two points (see Figure 3.2), and Ω_π interchanges these two points.

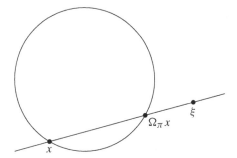

Fig. 3.1. Inversion Ω_π with spacelike pole ξ.

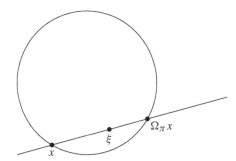

Fig. 3.2. Inversion Ω_π with timelike pole ξ.

Suppose, for example, we take $\xi = e_1$ and represent a point on Σ by homogeneous coordinates $(1, y)$, where y is a vector in the span of $\{e_2, \ldots, e_{n+2}\}$ satisfying the equation $y \cdot y = 1$, i.e., $y \in S^n$. Then from equation (3.16), we see that

$$\Omega_\pi(1, y) = (-1, y) \simeq (1, -y),$$

and Ω_π acts as the antipodal map on the sphere S^n. Note that the projective transformation Ω_π equals the projective transformation induced by the product $\Omega_2 \cdots \Omega_{n+2}$, where Ω_j is inversion in the polar hyperplane to e_j, because the corresponding orthogonal transformations differ by a minus sign. This is true independent of the choice of the hyperplane π with timelike pole. Therefore, we have established the following refinement of Theorem 3.8 for the Möbius group.

Theorem 3.9. *The Möbius group H is generated by inversions in polar hyperplanes to spacelike points in $\mathbf{P}^{n+1}$, i.e., by inversions in spheres in Σ.*

We now return to the setting of Lie's geometry of oriented spheres. In general, it is hard to give a geometric description of a Lie inversion that is not induced by a Möbius inversion. One noteworthy special case, however, is the change of orientation transformation Γ (see Remark 3.4) determined by the hyperplane π orthogonal to e_{n+3}.

We next present an alternative way to view the Lie sphere group G by decomposing it into certain natural subgroups. To do this, we need the concept of a linear complex of spheres. The *linear complex of spheres* determined by a point ξ in $\mathbf{P}^{n+2}$ is the set of all spheres represented by points x in the Lie quadric Q^{n+1} satisfying the equation $\langle x, \xi \rangle = 0$.

The complex is said to be *elliptic* if ξ is spacelike, *hyperbolic* if ξ is timelike, and *parabolic* if ξ is lightlike. Since the Lie sphere group G acts transitively on each of the three types of points, each linear complex of a given type looks like every other complex of the same type.

A typical example of an elliptic complex is obtained by taking $\xi = e_{n+2}$. A sphere S in $\mathbf{R}^n$ represented by a point x in Q^{n+1} satisfies the equation $\langle x, \xi \rangle = 0$ if and only if its coordinate $x_{n+2} = 0$ in $\mathbf{R}^n$, i.e., the center of S lies in the hyperplane $\mathbf{R}^{n-1}$ with equation $x_{n+2} = 0$ in $\mathbf{R}^n$. The linear complex consists of all spheres and planes orthogonal to this plane, including the points of the plane itself as a special case. A Lie sphere transformation T maps each sphere in the complex to another sphere in the complex if and only if $e_{n+2}^{\perp}$ is an invariant subspace of T. Since T can be represented by an orthogonal transformation, this is equivalent to $T[e_{n+2}] = [e_{n+2}]$. Thus T is determined by its action on $e_{n+2}^{\perp}$. Let $\mathbf{R}_2^{n+2}$ denote the vector subspace $e_{n+2}^{\perp}$ in $\mathbf{R}_2^{n+3}$ endowed with the metric $\langle, \rangle$ inherited from $\mathbf{R}_2^{n+3}$, and let $O(n, 2)$ denote the group of orthogonal transformations of the space $\mathbf{R}_2^{n+2}$. A transformation A in $O(n, 2)$ can be extended to $\mathbf{R}_2^{n+3}$ by setting $A e_{n+2} = e_{n+2}$. This gives an isomorphism between $O(n, 2)$ and the group of Lie sphere transformations which fix the elliptic complex. This group is a double covering of the group of Lie sphere transformations of the Euclidean space $\mathbf{R}^{n-1}$ orthogonal to e_{n+2} in $\mathbf{R}^n$.

A typical example of a hyperbolic complex is the case $\xi = e_{n+3}$. This complex consists of all point spheres. A second example is the complex corresponding to $\xi = (-r, r, 0, \ldots, 0, 1)$. This complex consists of all oriented spheres with signed radius r. The group of Lie sphere transformations which map this hyperbolic complex to itself consists of all transformations which map the projective point ξ to itself. This group is isomorphic to the Möbius subgroup of G, as discussed in Remark 3.4.

The parabolic complex determined by a point ξ in Q^{n+1} consists of all spheres in oriented contact with the sphere corresponding to ξ. A noteworthy example is the case $\xi = (1, -1, 0, \ldots, 0)$, the improper point. This system consists of all oriented hyperplanes in $\mathbf{R}^n$. A Lie sphere transformation which fixes this complex is called a *Laguerre transformation*, and the group of such Laguerre transformations is called the *Laguerre group*. We will study this group in detail in the next section.

3.4 Laguerre Geometry

Each point in the intersection of the Lie quadric Q^{n+1} with the plane

$$x_1 + x_2 = 0$$

represents either a plane in $\mathbf{R}^n$ or the improper point. The other points in the quadric represent actual spheres in $\mathbf{R}^n$ including point spheres. The homogeneous coordinates of points in this complementary set satisy the condition $x_1 + x_2 \neq 0$. The following elementary lemma shows that a Lie sphere transformation is determined by its action on such points.

Lemma 3.10. *A Lie sphere transformation is determined by its restriction to the set of points $[x]$ in Q^{n+1} with $x_1 + x_2 \neq 0$.*

Proof. To prove this, it is sufficient to exhibit a basis of lightlike vectors in $\mathbf{R}_2^{n+3}$ satisfying $x_1 + x_2 \neq 0$. One can check that $\{v_1, \ldots, v_{n+3}\}$ given below is such a basis:

$$v_1 = e_2 + e_{n+3}, \quad v_i = e_1 + e_i, \quad 2 \leq i \leq n+2, \quad v_{n+3} = e_3 - e_{n+3}. \qquad \square$$

We now show that the set of points in Q^{n+1} with $x_1 + x_2 \neq 0$ is naturally diffeomorphic to the Lorentz space $\mathbf{R}_1^{n+1}$ spanned by $\{e_3, \ldots, e_{n+3}\}$. By taking the appropriate scalar multiple, we may assume that the homogeneous coordinates of the point $[x]$ satisfy $x_1 + x_2 = 1$. Let

$$X = (x_3, \ldots, x_{n+3}),$$

and let

$$(X, X) = x_3^2 + \cdots + x_{n+2}^2 - x_{n+3}^2, \tag{3.17}$$

denote the restriction of $\langle , \rangle$ to $\mathbf{R}_1^{n+1}$. Then,

$$0 = \langle x, x \rangle = -x_1^2 + x_2^2 + (X, X) = -x_1 + x_2 + (X, X)$$

since $x_1 + x_2 = 1$. Hence we have $x_1 - x_2 = (X, X)$, and we can solve for x_1 and x_2 as follows:

$$x_1 = (1 + (X, X))/2, \quad x_2 = (1 - (X, X))/2. \tag{3.18}$$

Thus we have a diffeomorphism $[x] \mapsto X$ between the open set U of points in Q^{n+1} with $x_1 + x_2 \neq 0$ and points X in $\mathbf{R}_1^{n+1}$. In the proof of Theorem 3.5, it was shown that this diffeomorphism is conformal if Q^{n+1} is endowed with the standard pseudo-Riemannian metric (see Cahen and Kerbrat [14, p. 327]). From formula (2.14) of Chapter 2, p. 16, we see that the center p and signed radius r of the sphere determined by X are given by

$$p = (x_3, \ldots, x_{n+2}), \quad r = x_{n+3}. \tag{3.19}$$

Geometrically, one obtains the sphere S determined by X as the intersection of the plane $x_{n+3} = 0$ with the light (isotropy) cone with vertex X. The orientation of the sphere is determined by the sign of x_{n+3}. The mapping which takes X to the oriented sphere obtained this way was classically called *isotropy projection* (see Figure 3.3 and Blaschke [10, p. 136]).

Note that the spheres corresponding to two points X and Y in $\mathbf{R}_1^{n+1}$ are in oriented contact if and only if the line determined by X and Y is lightlike (see Figure 3.4).

To see this analytically, suppose that $x = (x_1, x_2, X)$ and $y = (y_1, y_2, Y)$ are the homogeneous coordinates of two points on the quadric satisfying

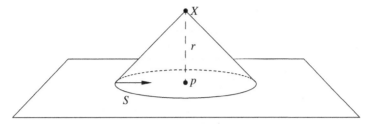

Fig. 3.3. Isotropy projection.

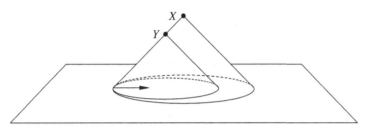

Fig. 3.4. Oriented contact and isotropy projection.

$$\langle x, x \rangle = \langle y, y \rangle = 0, \quad x_1 + x_2 = y_1 + y_2 = 1.$$

The spheres corresponding to $[x]$ and $[y]$ are in oriented contact if and only if $\langle x, y \rangle = 0$. Using equation (3.18) for x_1, x_2, y_1, y_2, a direct computation shows that

$$\langle x, y \rangle = -x_1 y_1 + x_2 y_2 + (X, Y) = -(X - Y, X - Y)/2. \tag{3.20}$$

Thus the spheres are in oriented contact if and only if the vector $X - Y$ is lightlike. From this, it is obvious that a parabolic pencil of spheres in $\mathbf{R}^n$ corresponds to a lightlike line in $\mathbf{R}_1^{n+1}$.

It is also possible to represent oriented hyperplanes of $\mathbf{R}^n$ in Laguerre geometry. The idea is to identify an oriented hyperplane π having unit normal N with the set of all contact elements (p, N), where p is a point of π. By isotropy projection, the parabolic pencil of spheres in oriented contact at (p, N) corresponds to a lightlike line in $\mathbf{R}_1^{n+1}$. The union of all these lightlike lines is an isotropy plane, i.e., an affine hyperplane V in $\mathbf{R}_1^{n+1}$ whose pole with respect to the Lorentz metric is lightlike. Such a plane meets $\mathbf{R}^n$ at an angle of $\pi/4$ (see Figure 3.5). Thus, we have a bijective correspondence between oriented planes in $\mathbf{R}^n$ and isotropy planes in $\mathbf{R}_1^{n+1}$.

A fundamental geometric quantity in Laguerre geometry is the tangential distance between two spheres. To study this, we first need to resolve the question of when two oriented spheres in $\mathbf{R}^n$ have a common tangent (oriented) plane. While it is obvious that some pairs of spheres have a common tangent plane (see Figure 3.6), it is just as obvious that some pairs, such as concentric spheres, do not. The following lemma answers this question.

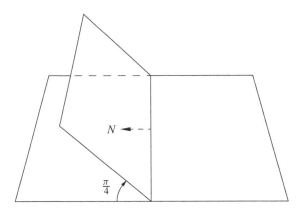

Fig. 3.5. Isotropy plane.

Lemma 3.11. *The two oriented spheres in $\mathbf{R}^n$ corresponding to the points X and Y in $\mathbf{R}^{n+1}_1$ have a common tangent plane if and only if $X - Y$ is lightlike or spacelike.*

Proof. From the discussion above, we know that the two spheres corresponding to X and Y have a common tangent plane if and only if there is an isotropy plane π in $\mathbf{R}^{n+1}_1$ that contains X and Y, i.e., there is a lightlike vector v such that $(X - Y, v) = 0$. First, suppose that $X - Y$ is timelike. Then the Lorentz metric is positive definite on the orthogonal complement of $X - Y$, so there is no nonzero lightlike vector v such that $(X - Y, v) = 0$, and thus there is no isotropy plane that contains X and Y. Next if $X - Y$ is lightlike, then the corresponding spheres are in oriented contact at a certain contact element (p, N), and they have exactly one common tangent plane determined by (p, N). Finally, suppose that $X - Y$ is spacelike. Then the Lorentz metric has signature $(n - 1, 1)$ on $(X - Y)^{\perp}$. Let v be any lightlike vector orthogonal to $X - Y$. Then the line $[X, Y]$ lies in the isotropy plane π through X with pole v, so the spheres determined by X and Y have a common tangent plane corresponding to the intersection of π with $\mathbf{R}^n$. This shows that the set of common tangent planes to two spheres is in bijective correspondence with the set of projective classes of lightlike vectors in the projective space $\mathbf{P}^{n-1}$ determined by the n-plane $(X - Y)^{\perp}$. As we saw in Section 2.1, this set is naturally diffeomorphic to an $(n - 2)$-sphere.   □

Suppose that $X = (p_1, r_1)$ and $Y = (p_2, r_2)$ are two points representing spheres S_1 and S_2 which have a common tangent plane (see Figure 3.6). By symmetry, it is clear that all common tangent segments to the two spheres have the same Euclidean length. This length d is called the *tangential distance* between S_1 and S_2. By constructing a right triangle as in Figure 3.6, we see that

$$d^2 + |r_1 - r_2|^2 = |p_1 - p_2|^2,$$

and so,

$$d^2 = |p_1 - p_2|^2 - |r_1 - r_2|^2 = (X - Y, X - Y). \tag{3.21}$$

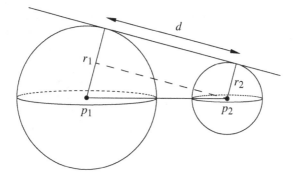

Fig. 3.6. Tangential distance between spheres.

Hence, the tangential distance is just the square root of the nonnegative quantity $(X - Y, X - Y)$. Of course, the tangential distance is zero precisely when the two spheres are in oriented contact.

A Lie sphere transformation that maps the improper point to itself is a Laguerre transformation. Since oriented contact must be preserved, Laguerre transformations can also be characterized as those Lie sphere transformations that take planes to planes (as noted in Section 3.3). Consequently, a Laguerre transformation maps the open set U determined by the condition $x_1 + x_2 \neq 0$ onto itself. Through the correspondence between U and $\mathbf{R}_1^{n+1}$ given via equation (3.18), a Laguerre transformation induces a transformation of $\mathbf{R}_1^{n+1}$ onto itself. We will now show that this must be an affine transformation.

Suppose that $\alpha = P(\sigma)$ is the Laguerre transformation determined by a transformation $\sigma \in O(n + 1, 2)$. As a transformation of $\mathbf{R}_2^{n+3}$, σ takes the affine plane π given by $x_1 + x_2 = 1$ to another affine plane in $\mathbf{R}_2^{n+3}$. If $\sigma\pi$ were not parallel to π, then $\sigma\pi$ would intersect the plane $x_1 + x_2 = 0$, contradicting the assumption that α maps U to U. Thus $\sigma\pi$ is given by the equation $x_1 + x_2 = c$, for some $c \neq 0$. If $A = \sigma/c$, then A induces the same Lie sphere transformation α as σ, but A takes the plane π to itself. Thus we now represent α by the transformation A, which is not in $O(n + 1, 2)$ unless $c = \pm 1$. Suppose that the transformation A is represented by the matrix

$$A = [a_{ij}], \quad 1 \leq i, j \leq n + 3,$$

with respect to the standard basis of $\mathbf{R}_2^{n+3}$. Since α takes the improper point to itself, we have

$$A(1, -1, 0, \ldots, 0) = \lambda(1, -1, 0, \ldots, 0),$$

for some $\lambda \neq 0$. From matrix multiplication and the equation above, we get

$$a_{i1} = a_{i2}, \quad 3 \leq i \leq n + 3. \tag{3.22}$$

Suppose now that $x = (x_1, \ldots, x_{n+3})$ with $x_1 + x_2 = 1$, and let

$$Ax = y = (y_1, \ldots, y_{n+3}).$$

By our choice of A, we know that $y_1 + y_2 = 1$ also. Thus x and y are determined by $X = (x_3, \ldots, x_{n+3})$ and $Y = (y_3, \ldots, y_{n+3})$ using formula (3.18). Since $a_{i1} = a_{i2}$ and $x_1 + x_2 = 1$, we have

$$y_i = a_{i1}x_1 + a_{i2}x_2 + \sum_{j=3}^{n+3} a_{ij}x_j = a_{i1} + \sum_{j=3}^{n+3} a_{ij}x_j. \tag{3.23}$$

Thus, Y is obtained from X by the affine transformation T given by

$$Y = TX = BX + C, \tag{3.24}$$

where B is the invertible linear transformation of $\mathbf{R}_1^{n+1}$ represented by the matrix $[a_{ij}]$, $3 \leq i, j \leq n + 3$, and C is the vector $(a_{31}, \ldots, a_{(n+3)1})$. The fact that this transformation must preserve oriented contact of spheres implies further that T must preserve the relationship

$$(X - Z, X - Z) = 0. \tag{3.25}$$

The discussion above suggests the possibility of simply beginning in the space $\mathbf{R}_1^{n+1}$, as in Blaschke [10, p. 136]. From that point of view, an affine transformation $TX = BX + C$ of $\mathbf{R}_1^{n+1}$ which preserves the relationship (3.25) is called an *affine Laguerre transformation*. If the vector C is zero, then T is called a *linear Laguerre transformation*. Of course, the linear part B must be invertible by definition of an affine transformation.

Remark 3.12. Blaschke [10, p. 141] referred to the group of affine Laguerre transformations as the "extended Laguerre group." He called subgroup of transformations with the property that the linear part B is in $O(n, 1)$ the "restricted Laguerre group" or simply the "Laguerre group." From equation (3.25), we see that these transformations can also be characterized as affine Laguerre transformations that preserve tangential distances between spheres. This restricted Laguerre group is sometimes used as the basis for Laguerre geometry (see Blaschke [10, pp. 268–272], Li [102]).

We have seen that a Lie sphere transformation that takes the improper point to itself induces an affine Laguerre transformation. Conversely, we will show that an affine Laguerre transformation extends in a unique way to a Lie sphere transformation. Before that, however, we want to establish some important properties of the group of Laguerre transformations. Theorem 3.1 and the assumption that equation (3.25) is preserved yield the following result concerning the linear part of an affine Laguerre transformation.

Theorem 3.13. *Suppose that* $TX = BX + C$ *is an affine Laguerre transformation. Then* $B = \mu D$, *where* $D \in O(n, 1)$ *and* $\mu > 0$.

Proof. Suppose that X and Z are any two points in $\mathbf{R}_1^{n+1}$ such that $X - Z$ is lightlike. Then, since T preserves equation (3.25), we have

$$0 = ((BX + C) - (BZ + C), (BX + C) - (BZ + C)) = (B(X - Z), B(X - Z)).$$

Thus the linear transformation B takes lightlike vectors to lightlike vectors. By Theorem 3.1 and our standing assumption that $n \geq 2$, we know that there exists a positive number λ such that $(Bv, Bw) = \lambda(v, w)$, for all v, w in $\mathbf{R}_1^{n+1}$. Then $D = \lambda^{-1/2} B$ is orthogonal, and we have $B = \mu D$ for $\mu = \lambda^{1/2}$. $\qquad\square$

We now want to give a geometric interpretation of this decomposition of B. First, it is immediate from equation (3.21) and Theorem 3.13 that the transformations in $O(n, 1)$ are precisely those linear transformations which preserve tangential distances between spheres. Secondly, consider the linear transformation $S_\mu = \mu I$, where $\mu > 0$ and I is the identity transformation on $\mathbf{R}_1^{n+1}$. This transformation takes the point (p, r) to $(\mu p, \mu r)$. When interpreted as a map on the space of spheres, it takes a sphere with center p and signed radius r to the sphere with center μp and signed radius μr. Thus S_μ is one of the two affine Laguerre transformations induced from the Euclidean central dilatation $p \mapsto \mu p$, for $p \in \mathbf{R}^n$. The transformation S_μ preserves the sign of the radius and hence the orientation of each sphere in $\mathbf{R}^n$. The other affine Laguerre transformation induced from the same central dilatation is ΓS_μ, where Γ is the change of orientation transformation. Thus we have the following geometric version of Theorem 3.13.

Corollary 3.14. *Every linear Laguerre transformation B can be written in the form $B = S_\mu D$, where S_μ is the orientation preserving Laguerre transformation induced by a central dilatation of $\mathbf{R}^n$, and D preserves tangential distances between spheres.*

Next we consider the effect of a *Laguerre translation*, $TX = X + C$, on the space of spheres. Suppose first that $C = (v, 0)$ for $v \in \mathbf{R}^n$. Then we have

$$T(p, r) = (p + v, r),$$

so T translates the center of every sphere by the vector v while preserving the signed radius. We see that T is just the orientation preserving affine Laguerre transformation induced from the Euclidean translation $p \mapsto p + v$. We denote T by τ_v.

Now consider the case where $C = (0, t), t \in \mathbf{R}$. Then $T(p, r) = (p, r + t)$, so T adds t to the signed radius of every sphere while keeping the center fixed. T is called *parallel transformation* by t and will be denoted P_t. Note that P_t is a Laguerre transformation that is not a Möbius transformation. It takes point spheres to spheres with signed radius t and takes spheres of radius $-t$ to point spheres. Thus, the group of Laguerre translations is a commutative subgroup of the group of affine Laguerre transformations, and we have shown that it decomposes as follows.

Theorem 3.15. *Any Laguerre translation T can be written in the form*

$$T = P_t \tau_v,$$

where P_t is a parallel transformation, and τ_v is the orientation preserving Laguerre translation induced by Euclidean translation by the vector v.

In studying the role of inversions in the group of affine Laguerre transformations, we must consider inversions in planes in $\mathbf{R}_1^{n+1}$ which do not contain the origin. Let π be an affine plane in $\mathbf{R}_1^{n+1}$ whose pole ξ is not lightlike. Then the inversion Ω_π of $\mathbf{R}_1^{n+1}$ in π is defined by the formula

$$\Omega_\pi X = X - \frac{2(X - P, \xi)}{(\xi, \xi)}\xi, \tag{3.26}$$

where P is any point on the plane π. We will call Ω_π a *Laguerre inversion*. It is easy to see that the group of affine Laguerre transformations is generated by Laguerre inversions along with transformation of the form S_μ induced by central dilatations. First, we know from Theorem 3.13 that any linear Laguerre transformation is the product of some S_μ with an orthogonal transformation D. An orthogonal transformation D is a product of Laguerre inversions by Theorem 3.7. As for translations, it is well known that Euclidean translation by a vector v is a product of two Euclidean reflections $\Omega_2\Omega_1$ in parallel planes π_1 and π_2 orthogonal to v such that $v/2$ is the vector from any given point P on π_1 to its closest point Q on π_2. (See [165, p. 23] for more detail.) The Laguerre translation τ_v is the product of the Laguerre inversions induced by these two Euclidean reflections.

To express a parallel transformation P_t as the product of two inversions, we first note that the change of orientation transformation Γ is a Laguerre inversion in the plane π_1 through the origin with pole e_{n+3}. If Ω_2 is the Laguerre inversion in the plane π_2 through the point $(0, t/2)$ with pole e_{n+3}, then P_t is the product $\Omega_2\Gamma$.

Finally, we want to show that every affine Laguerre transformationextends to a Lie sphere transformation. Such an extension is necessarily unique by Lemma 3.10. First, recall that if α is the Lie sphere transformation extending an affine Laguerre transformation $TX = BX + C$, then α has a unique representative A in $GL(n + 3)$ which takes the plane $x_1 + x_2 = 1$ to itself. The matrix $[a_{ij}]$ for A with respect to the standard basis is largely determined by B and C through equation (3.24), i.e.,

$$[B] = [a_{ij}], \quad 3 \le i, j \le n + 3, \quad C = (a_{31}, \ldots, a_{(n+3)1}). \tag{3.27}$$

We now show how to determine the rest of A.

First consider the case where B is a linear Laguerre transformation. The Lie extension of B must take the improper point $[e_1 - e_2]$ to itself and the point $[e_1 + e_2]$ corresponding to the origin in $\mathbf{R}_1^{n+1}$ to itself. This means that the transformation A satisfies

$$A(e_1 - e_2) = a(e_1 - e_2), \quad A(e_1 + e_2) = b(e_1 + e_2), \tag{3.28}$$

for some nonzero scalars a and b. From equation (3.28) and the linearity of A, we obtain

$$Ae_1 = ce_1 + de_2, \quad Ae_2 = de_1 + ce_2, \tag{3.29}$$

where $c = (a+b)/2$ and $d = (b-a)/2$. Thus, the span $\mathbf{R}_1^2$ of e_1 and e_2 is an invariant subspace of A. Since A is a scalar multiple of an orthogonal transformation, the orthogonal complement $\mathbf{R}_1^{n+1}$ of $\mathbf{R}_1^2$ is also invariant under A. Therefore, the matrix A has the form

$$A = \begin{bmatrix} J & 0 \\ 0 & B \end{bmatrix} \quad \text{where } J = \begin{bmatrix} c & d \\ d & c \end{bmatrix}. \tag{3.30}$$

By Corollary 3.14, every linear Laguerre transformation is of the form $S_\mu D$, where S_μ is induced from a central dilatation of $\mathbf{R}^n$ and $D \in O(n, 1)$. We now show how to extend each of these two types of transformations. First, consider the case of a linear Laguerre transformation determined by $D \in O(n, 1)$. The Lie extension of D is obtained by taking $c = 1, d = 0$ and $B = D$ in equation (3.30). To check this, let X be an arbitrary point in $\mathbf{R}_1^{n+1}$. By equation (3.18), X corresponds to the point in Q^{n+1} with homogeneous coordinates

$$x = (1 + (X, X), 1 - (X, X), 2X)/2. \tag{3.31}$$

The point DX in $\mathbf{R}_1^{n+1}$ corresponds to the point in Q^{n+1} with homogeneous coordinates

$$y = (1 + (DX, DX), 1 - (DX, DX), 2DX)/2.$$

However, since $(DX, DX) = (X, X)$, we have

$$y = (1 + (X, X), 1 - (X, X), 2DX)/2.$$

It is now clear that $y = Ax$, where A is in the form of equation (3.30) with $c = 1, d = 0$ and $B = D$. Note that this matrix is in $O(n + 1, 2)$.

Next consider the linear Laguerre transformation $S_\mu = \mu I$ on $\mathbf{R}_1^{n+1}$. Let A have the form of equation (3.30) with $B = \mu I$. Let $X \in \mathbf{R}_1^{n+1}$ and let x be as in equation (3.31). By matrix multiplication, the first two coordinates of $y = Ax$ are given by

$$y_1 = ((c + d) + (c - d)(X, X))/2, \quad y_2 = ((c + d) + (d - c)(X, X))/2. \tag{3.32}$$

On the other hand, the point μX corresponds to the point in Q^{n+1} with homogeneous coordinates

$$(1 + \mu^2(X, X), 1 - \mu^2(X, X), 2\mu X)/2. \tag{3.33}$$

Equating y_1 and y_2 with the first two coordinates in equation (3.33) yields

$$c = (1 + \mu^2)/2, \quad d = (1 - \mu^2)/2. \tag{3.34}$$

The matrix A in equation (3.30) with these values of c and d and $B = \mu I$ is the extension of S_μ. Note that the numbers c and d in equation (3.34) satisfy $c^2 - d^2 = \mu^2$. Since $\mu > 0$, the number c/μ is positive, and so there exists a real number t such that

$$c/\mu = \cosh t, \quad d/\mu = \sinh t. \tag{3.35}$$

From equations (3.34) and (3.35) one can determine that $t = -\log \mu$. If we divide the matrix A by μ, we obtain the following orthogonal matrix which also represents the Lie sphere transformation extending S_μ,

$$\begin{bmatrix} K & 0 \\ 0 & I \end{bmatrix} \quad \text{where } K = \begin{bmatrix} \cosh t & \sinh t \\ \sinh t & \cosh t \end{bmatrix}. \tag{3.36}$$

Finally, we turn to the problem of extending the parallel transformation P_t. The linear part B of P_t is the identity I, while the translation part is the vector $C = (0, t)$. By equation (3.27), this determines much of the matrix A of the extension of P_t. Further, since the improper point is mapped to itself,

$$A(1, -1, 0, \ldots, 0) = a(1, -1, 0, \ldots, 0), \tag{3.37}$$

for some $a \neq 0$. Since P_t takes the point $(0, 0)$ in $\mathbf{R}_1^{n+1}$ to $(0, t)$, the transformation A must satisfy

$$A(1, 1, 0, \ldots, 0) = b(1 - t^2, 1 + t^2, 0, 2t)/2, \tag{3.38}$$

for some $b \neq 0$. By adding equations (3.37) and (3.38), we can determine Ae_1. Since $a_{(n+3)1} = t$ by equation (3.27), b must equal 2. From the results so far, we know that the subspace,

$$V = \mathrm{Span}\{e_1, e_2, e_{n+3}\},$$

is invariant under A. Therefore $V^\perp$ is also invariant. This and the fact that $B = I$ imply that $Ae_i = e_i$, for $3 \leq i \leq n + 2$. Since A magnifies all vector lengths by the same amount, this implies that A is orthogonal. From these orthogonal relations, one can determine that $a = 1$ and ultimately that A has the form below. For future reference, we will now denote this transformation by P_t instead of A, so we have shown

$$P_t = \begin{bmatrix} 1 - (t^2/2) & -t^2/2 & 0 \ldots 0 & -t \\ t^2/2 & 1 + (t^2/2) & 0 \ldots 0 & t \\ 0 & 0 & I & 0 \\ t & t & 0 \ldots 0 & 1 \end{bmatrix}. \tag{3.39}$$

3.5 Subgeometries of Lie Sphere Geometry

We close this chapter by examining some important subgeometries of Lie sphere geometry from the point of view of Klein's Erlangen Program. These are the geometries of Möbius and Laguerre, and the Euclidean, spherical and hyperbolic metric geometries. By making use of the concept of a Legendre submanifold, to be introduced in the next chapter, one can study submanifold theory in any of these subgeometries within the context of Lie sphere geometry.

The subgroup of Möbius transformations consists of those Lie sphere transformations that map point spheres to point spheres. These are precisely the Lie sphere transformations that map the point $[e_{n+3}]$ to itself. As we saw in Remark 3.4, this Möbius group is isomorphic to $O(n + 1, 1)$.

The subgroup of Laguerre transformations consists of those Lie sphere transformations that map hyperplanes to hyperplanes in $\mathbf{R}^n$. These are the Lie sphere transformations that map the improper point $[e_1 - e_2]$ to itself. In the preceding section, we saw that each Laguerre transformation corresponds to an affine Laguerre transformation of the space $\mathbf{R}_1^{n+1}$ spanned by $\{e_3, \ldots, e_{n+3}\}$.

As before let $\mathbf{R}^n$ denote the Euclidean space spanned by the vectors $\{e_3, \ldots, e_{n+2}\}$. Recall that a *similarity transformation* of $\mathbf{R}^n$ is a mapping ϕ from $\mathbf{R}^n$ to itself, such that for all p and q in $\mathbf{R}^n$, the Euclidean distance $d(p, q)$ is transformed as follows:

$$d(\phi p, \phi q) = \kappa d(p, q),$$

for some constant $\kappa > 0$. Every similarity transformation can be written as a central dilatation followed by an isometry of $\mathbf{R}^n$. The group of Lie sphere transformations induced by similarity transformations is clearly a subgroup of both the Laguerre group and the Möbius group. The next theorem shows that it is precisely the intersection of these two subgroups.

Theorem 3.16.
(a) *The intersection of the Laguerre group and the Möbius group is the group of Lie sphere transformations induced by similarity transformations of $\mathbf{R}^n$.*
(b) *The group G of Lie sphere transformations is generated by the union of the groups of Laguerre and Möbius.*

Proof.
(a) By the results of the last section, the Laguerre group is isomorphic to the group of affine Laguerre transformations on $\mathbf{R}_1^{n+1}$. An affine Laguerre transformation $TX = BX + C$ is also a Möbius transformation if and only if

$$T e_{n+3} = \pm e_{n+3}.$$

Since $T(0) = C$, this immediately implies that $C = (v, 0)$, for some vector $v \in \mathbf{R}^n$. Next by Corollary 3.14, the linear part B of T is of the form $S_\mu D$, where S_μ is induced from a central dilatation of $\mathbf{R}^n$ and D is in $O(n, 1)$. Since S_μ is a Möbius transformation, T is a Möbius transformation precisely when D is a Möbius transformation, i.e., $D e_{n+3} = \pm e_{n+3}$. This means that the matrix for D with respect to the standard basis of $\mathbf{R}_1^{n+1}$ has the form

$$D = \begin{bmatrix} A & 0 \\ 0 & \pm 1 \end{bmatrix}, \quad A \in O(n). \tag{3.40}$$

Thus, D is one of the two Laguerre transformations induced by the linear isometry A of $\mathbf{R}^n$, and B is a similarity transformation. Then T is also a similarity transformation, and (a) is proven.

(b) Let α be a Lie sphere transformation. If $\alpha[e_1 - e_2] = [e_1 - e_2]$, then α is a Laguerre transformation. Next suppose that $\alpha[e_1 - e_2]$ is a point $[x]$ in Q^{n+1} with $x_1 + x_2 = 0$. Then $\alpha[e_1 - e_2]$ corresponds to a plane π in $\mathbf{R}^n$. Let I_1 be an inversion in a sphere whose center is not on the plane π. Then $[y] = I_1[x]$ is a point in Q^{n+1} with $y_1 + y_2 \neq 0$, i.e., $[y]$ corresponds to a sphere in $\mathbf{R}^n$. Let p and r denote the center and signed radius of this sphere. If $\alpha[e_1 - e_2]$ does not correspond to a plane, then the step above is not needed. In that case, we let I_1 be the identity transformation.

Next, the parallel transformation P_{-r} takes the point $[y]$ to the point $[z]$ with $z_{n+3} = 0$ corresponding to the point sphere p in $\mathbf{R}^n$. Finally, an inversion I_2 in a sphere centered at p takes $[z]$ to the improper point $[e_1 - e_2]$. Since the transformation $I_2 P_{-r} I_1 \alpha$ takes $[e_1 - e_2]$ to itself, it is a Laguerre transformation ψ. Since each inversion is its own inverse and the inverse of P_{-r} is P_r, we have $\alpha = I_1 P_r I_2 \psi$, a product of Laguerre and Möbius transformations.   □

From the proof of part (a) of Theorem 3.16, we have the following immediate corollary.

Corollary 3.17. *The group of Lie sphere transformations induced from isometries of* $\mathbf{R}^n$ *is isomorphic to the set of affine Laguerre transformations* $TX = DX + C$, *where* D *has the form* (3.40) *and* $C = (v, 0)$, *for* $v \in \mathbf{R}^n$.

Two other important subgeometries of Möbius geometry are the geometries of the sphere S^n and hyperbolic space H^n (see Section 2.4). Let S^n be the unit sphere in the Euclidean space $\mathbf{R}^{n+1}$ spanned by $\{e_2, \ldots, e_{n+2}\}$. The group of isometries of S^n is the orthogonal group $O(n + 1)$ of linear transformations that preserve the metric on $\mathbf{R}^{n+1}$. An isometry A of S^n induces a Möbius transformation whose matrix with respect to the standard basis $\{e_1, \ldots, e_{n+2}\}$ of $\mathbf{R}_1^{n+2}$ is

$$\begin{bmatrix} 1 & 0 \\ 0 & A \end{bmatrix}. \tag{3.41}$$

The group of isometries of S^n is clearly isomorphic to the subgroup of Möbius transformations of this form. Of course, each Möbius transformation induces two Lie sphere transformations differing by the change of orientation transformation Γ.

As in Section 2.4, we represent hyperbolic space H^n by the set of points y in Lorentz space $\mathbf{R}_1^{n+1}$ spanned by $\{e_1, e_3, \ldots, e_{n+2}\}$ satisfying $(y, y) = -1$, where $y_1 \geq 1$. (Note that this is a different Lorentz space than the one used in Laguerre geometry.) H^n is one of the two components of the subset of $\mathbf{R}_1^{n+1}$ determined by the equation $(y, y) = -1$. The group of isometries of H^n is a subgroup of index 2 in the orthogonal group $O(n, 1)$ of linear isometries of $\mathbf{R}_1^{n+1}$, since precisely one of the two orthogonal transformations A and $-A$ takes the component H^n to itself. There are two ways to extend an isometry A of H^n to an orthogonal transformation B of

$$\mathbf{R}_1^{n+2} = \text{Span}\{e_1, \ldots, e_{n+2}\};$$

namely, one can define Be_2 to be e_2 or $-e_2$. These two extensions induce different Möbius transformations, since they do not differ by a sign. On the other hand, the extension determined by setting $Be_2 = -e_2$ is projectively equivalent to the extension C of $-A$ satisfying $Ce_2 = e_2$. Hence, the group of Möbius transformations induced from isometries of H^n is isomorphic to $O(n, 1)$ itself. As before, each of these transformations induces two Lie sphere transformations.

For both the spherical and hyperbolic metrics, there is a parallel transformation P_t that adds t to the signed radius of each sphere while keeping the center fixed. As we saw in Section 2.4, the sphere in S^n with center p and signed radius ρ is represented by the point $[(\cos \rho, p, \sin \rho)]$ in Q^{n+1}. One easily checks that *spherical parallel transformation* P_t is accomplished by the following transformation in $O(n + 1, 2)$:

$$P_t e_1 = \cos t \, e_1 + \sin t \, e_{n+3},$$
$$P_t e_{n+3} = -\sin t \, e_1 + \cos t \, e_{n+3}, \qquad (3.42)$$
$$P_t e_i = e_i, \quad 2 \le i \le n+2.$$

In H^n the sphere with center $p \in H^n$ and signed radius ρ corresponds to the point $[p + \cosh \rho \, e_2 + \sinh \rho \, e_{n+3}]$ in Q^{n+1}. Thus *hyperbolic parallel transformation* is accomplished by the following transformation:

$$P_t e_i = e_i, \quad i = 1, 3, \ldots, n+2.$$
$$P_t e_2 = \cosh t \, e_2 + \sinh t \, e_{n+3}, \qquad (3.43)$$
$$P_t e_{n+3} = \sinh t \, e_2 + \cosh t \, e_{n+3}.$$

The following theorem demonstrates the important role played by parallel transformations in generating the Lie sphere group (see Cecil–Chern [37]).

Theorem 3.18. *Any Lie sphere transformation α can be written as*

$$\alpha = \phi P_t \psi,$$

where ϕ and ψ are Möbius transformations and P_t is some Euclidean, spherical or hyperbolic parallel transformation.

Proof. Represent α by a transformation $A \in O(n+1, 2)$. If $A e_{n+3} = \pm e_{n+3}$, then α is a Möbius transformation. If not, then $A e_{n+3}$ is some unit timelike vector v linearly independent from e_{n+3}. The plane $[e_{n+3}, v]$ in $\mathbf{R}_2^{n+3}$ can have signature $(-, -)$, $(-, +)$ or $(-, 0)$. In the case where the plane has signature $(-, -)$, we can write

$$v = -\sin t \, u_1 + \cos t \, e_{n+3},$$

where u_1 is a unit timelike vector orthogonal to e_{n+3}, and $0 < t < \pi$. Let ϕ be a Möbius transformation such that $\phi^{-1} u_1 = e_1$. Then from equation (3.42), we see that $P_{-t} \phi^{-1} v = e_{n+3}$. Hence,

$$P_{-t} \phi^{-1} \alpha e_{n+3} = e_{n+3},$$

i.e., $P_{-t} \phi^{-1} \alpha$ is a Möbius transformation ψ. Thus, $\alpha = \phi P_t \psi$, as desired.

The other two cases are similar. If the plane $[e_{n+3}, v]$ has signature $(-, 0)$, then we can write

$$v = -t u_1 + t u_2 + e_{n+3},$$

where u_1 and u_2 are unit timelike and spacelike vectors, respectively, orthogonal to e_{n+3} and to each other. If ϕ is a Möbius transformation such that $\phi^{-1} u_1 = e_1$ and $\phi^{-1} u_2 = e_2$, then $P_{-t} \phi \alpha$ is a Möbius transformation ψ, where P_t is the Euclidean parallel transformation given in equation (3.39). As before, we get $\alpha = \phi P_t \psi$. Finally, if the plane $[e_{n+3}, v]$ has signature $(-, +)$, then

$$v = \sinh t \, u_2 + \cosh t \, e_{n+3},$$

for a unit spacelike vector u_2 orthogonal to e_{n+3}. Let ϕ be a Möbius transformation such that $\phi^{-1} u_2 = e_2$, and conclude that $\alpha = \phi P_t \psi$ for the hyperbolic parallel transformation P_t in equation (3.43).   □

4

Legendre Submanifolds

In this chapter, we develop the framework necessary to study submanifolds within the context of Lie sphere geometry. The manifold Λ^{2n-1} of projective lines on the Lie quadric Q^{n+1} has a contact structure, i.e., a globally defined 1-form ω such that $\omega \wedge (d\omega)^{n-1} \neq 0$ on Λ^{2n-1}. This gives rise to a codimension one distribution D on Λ^{2n-1} that has integral submanifolds of dimension $n-1$, but none of higher dimension. These integral submanifolds are called Legendre submanifolds. Any submanifold of a real space-form $\mathbf{R}^n$, S^n or H^n naturally induces a Legendre submanifold, and thus Lie sphere geometry can be used to analyze submanifolds in these spaces. This has been particularly effective in the classification of Dupin submanifolds, which are defined in Section 4.4. In Section 4.5, we define the Lie curvatures of a Legendre submanifold. These are natural Lie invariants which have proved to be valuable in the study of Dupin submanifolds but are defined on the larger class of Legendre submanifolds. We then give a Lie geometric characterization of those Legendre submanifolds which are Lie equivalent to an isoparametric hypersurface in a sphere (Theorem 4.19). In Section 4.6, we prove that tautness is invariant under Lie sphere transformations. In Section 4.7, we discuss the construction of Ferus, Karcher, and Münzner [73] of isoparametric hypersurfaces with four principal curvatures, based on representations of Clifford algebras. Finally, in Section 4.8, we discuss the counterexamples of Pinkall–Thorbergsson and Miyaoka–Ozawa to the conjecture that a compact proper Dupin hypersurface in a sphere must be Lie equivalent to an isoparametric hypersurface. For a treatment of submanifold theory in Möbius geometry, see Blaschke [10], C.-P. Wang [194]–[196], or Hertrich–Jeromin [89], and in Laguerre geometry, see Blaschke [10], Musso and Nicolodi [124]–[125], Li [102], or Li and Wang [103].

4.1 Contact Structure on Λ^{2n-1}

In this section, we demonstrate explicitly that the manifold Λ^{2n-1} of projective lines on the Lie quadric Q^{n+1} is a contact manifold. The reader is referred to Arnold [3, p. 349] or Blair [9] for a more complete treatment of contact manifolds in general.

As in earlier chapters, let $\{e_1, \ldots, e_{n+3}\}$ denote the standard orthonormal basis for $\mathbf{R}_2^{n+3}$ with e_1 and e_{n+3} timelike. We consider S^n to be the unit sphere in the Euclidean space $\mathbf{R}^{n+1}$ spanned by $\{e_2, \ldots, e_{n+2}\}$. A *contact element* on S^n is a pair (x, ξ), where $x \in S^n$ and ξ is a unit tangent vector to S^n at x. Thus, the space of contact elements is the unit tangent bundle $T_1 S^n$. We consider $T_1 S^n$ to be the $(2n - 1)$-dimensional submanifold of $S^n \times S^n \subset \mathbf{R}^{n+1} \times \mathbf{R}^{n+1}$ given by

$$T_1 S^n = \{(x, \xi) \mid \quad |x| = 1, \quad |\xi| = 1, \quad x \cdot \xi = 0\}. \tag{4.1}$$

In general, a $(2n - 1)$-dimensional manifold V^{2n-1} is said to be a *contact manifold* if it carries a global 1-form ω such that

$$\omega \wedge (d\omega)^{n-1} \neq 0 \tag{4.2}$$

at all points of V^{2n-1}. Such a form ω is called a *contact form*. It is known (see, for example, [9, p. 10]) that the unit tangent bundle $T_1 M$ of any n-dimensional Riemannian manifold M is a $(2n - 1)$-dimensional contact manifold. A contact form ω defines a codimension one distribution D on V^{2n-1},

$$D_p = \{Y \in T_p V^{2n-1} \mid \omega(Y) = 0\}, \tag{4.3}$$

for $p \in V^{2n-1}$, called the *contact distribution*. This distribution is as far from being integrable as possible, in that there exist integral submanifolds of D of dimension $n - 1$ but none of higher dimension (see Theorem 4.1 below). A contact distribution determines the corresponding contact form up to multiplication by a nonvanishing smooth function.

In our particular case, a tangent vector to $T_1 S^n$ at a point (x, ξ) can be written in the form (X, Z) where

$$X \cdot x = 0, \quad Z \cdot \xi = 0. \tag{4.4}$$

Differentiation of the condition $x \cdot \xi = 0$ implies that (X, Z) must also satisfy

$$X \cdot \xi + Z \cdot x = 0. \tag{4.5}$$

We will show that the form ω defined by

$$\omega(X, Z) = X \cdot \xi, \tag{4.6}$$

is a contact form on $T_1 S^n$. Thus, at a point (x, ξ), the distribution D is the $(2n - 2)$-dimensional space of vectors (X, Z) satisfying $X \cdot \xi = 0$, as well as the equations (4.4) and (4.5). Of course, the equation $X \cdot \xi = 0$ together with equation (4.5) implies that

$$Z \cdot x = 0, \tag{4.7}$$

for vectors (X, Z) in D. To see that ω satisfies the condition (4.2), we will identify $T_1 S^n$ with the manifold Λ^{2n-1} of projective lines on the Lie quadric Q^{n+1} and compute $d\omega$ using the method of moving frames. The results in this calculation will turn out to be useful in our general study of submanifolds.

We establish a bijective correspondence between the points of $T_1 S^n$ and the lines on Q^{n+1} by the map

$$(x, \xi) \mapsto [Y_1(x, \xi), Y_{n+3}(x, \xi)], \tag{4.8}$$

where

$$Y_1(x, \xi) = (1, x, 0), \qquad Y_{n+3}(x, \xi) = (0, \xi, 1). \tag{4.9}$$

The points on a line on Q^{n+1} correspond to a parabolic pencil of spheres in S^n. By formula (2.21) of Chapter 2, p. 17, the point $[Y_1(x, \xi)]$ corresponds to the unique point sphere in the pencil determined by the line $[Y_1(x, \xi), Y_{n+3}(x, \xi)]$, and $Y_{n+3}(x, \xi)$ corresponds to the unique great sphere in the pencil. Since every line on the quadric contains exactly one point sphere and one great sphere by Corollary 2.7 of Chapter 2, p. 22, the correspondence in (4.8) is bijective. We use the correspondence (4.8) to put a differentiable structure on the manifold Λ^{2n-1} in such a way that the map in (4.8) becomes a diffeomorphism.

We now introduce the method of moving frames in the context of Lie sphere geometry, as in Cecil–Chern [37]. The reader is also referred to Cartan [15], Griffiths [78], Jensen [93] or Spivak [171, Vol. 2, Chapter 7] for an exposition of the general method. Since we want to define frames on the manifold Λ^{2n-1}, it is better to use frames for which some of the vectors are lightlike, rather than orthonormal frames. To facilitate the exposition, we will use the following range of indices in this section:

$$1 \le a, b, c \le n + 3, \quad 3 \le i, j, k \le n + 1. \tag{4.10}$$

A *Lie frame* is an ordered set of vectors $\{Y_1, \ldots, Y_{n+3}\}$ in $\mathbf{R}_2^{n+3}$ satisfying the relations

$$\langle Y_a, Y_b \rangle = g_{ab}, \tag{4.11}$$

for

$$[g_{ab}] = \begin{bmatrix} J & 0 & 0 \\ 0 & I_{n-1} & 0 \\ 0 & 0 & J \end{bmatrix}, \tag{4.12}$$

where I_{n-1} is the $(n-1) \times (n-1)$ identity matrix and

$$J = \begin{bmatrix} 0 & 1 \\ 1 & 0 \end{bmatrix}. \tag{4.13}$$

If $(y_1, \ldots, y_{n+3})$ are homogeneous coordinates on $\mathbf{P}^{n+2}$ with respect to a Lie frame, then the Lie metric has the form

$$\langle y, y \rangle = 2(y_1 y_2 + y_{n+2} y_{n+3}) + y_3^2 + \cdots + y_{n+1}^2. \tag{4.14}$$

The space of all Lie frames can be identified with the group $O(n+1, 2)$ of which the Lie sphere group G, being isomorphic to $O(n+1, 2)/\{\pm I\}$, is a quotient group. In this space, we introduce the *Maurer–Cartan forms* ω_a^b by the equation

$$dY_a = \sum \omega_a^b Y_b, \tag{4.15}$$

and we adopt the convention that the sum is always over the repeated index. Differentiating equation (4.11), we get

$$\omega_{ab} + \omega_{ba} = 0, \tag{4.16}$$

where

$$\omega_{ab} = \sum g_{bc} \omega_a^c. \tag{4.17}$$

Equation (4.16) says that the following matrix is skew-symmetric,

$$[\omega_{ab}] = \begin{bmatrix} \omega_1^2 & \omega_1^1 & \omega_1^i & \omega_1^{n+3} & \omega_1^{n+2} \\ \omega_2^2 & \omega_2^1 & \omega_2^i & \omega_2^{n+3} & \omega_2^{n+2} \\ \omega_j^2 & \omega_j^1 & \omega_j^i & \omega_j^{n+3} & \omega_j^{n+2} \\ \omega_{n+2}^2 & \omega_{n+2}^1 & \omega_{n+2}^i & \omega_{n+2}^{n+3} & \omega_{n+2}^{n+2} \\ \omega_{n+3}^2 & \omega_{n+3}^1 & \omega_{n+3}^i & \omega_{n+3}^{n+3} & \omega_{n+3}^{n+2} \end{bmatrix}. \tag{4.18}$$

Taking the exterior derivative of equation (4.15) yields the *Maurer–Cartan equations*,

$$d\omega_a^b = \sum \omega_a^c \wedge \omega_c^b. \tag{4.19}$$

We now produce a contact form on $T_1 S^n$ in the context of moving frames. We want to choose a local frame $\{Y_1, \ldots, Y_{n+3}\}$ on $T_1 S^n$ with Y_1 and Y_{n+3} given by equation (4.9). When we transfer this frame to Λ^{2n-1}, it will have the property that for each point $\lambda \in \Lambda^{2n-1}$, the line $[Y_1, Y_{n+3}]$ of the frame λ is the line on the quadric Q^{n+1} corresponding to λ.

On a sufficiently small open subset U in $T_1 S^n$, we can find smooth mappings,

$$v_i : U \to \mathbf{R}^{n+1}, \quad 3 \le i \le n+1,$$

such that at each point $(x, \xi) \in U$, the vectors $v_3(x, \xi), \ldots, v_{n+1}(x, \xi)$ are unit vectors orthogonal to each other and to x and ξ. By equations (4.4) and (4.5), we see that the vectors

$$\{(v_i, 0), (0, v_i), (\xi, -x)\}, \quad 3 \le i \le n+1, \tag{4.20}$$

form a basis to the tangent space to $T_1 S^n$ at (x, ξ). We now define a Lie frame on U as follows:

$$\begin{aligned} Y_1(x, \xi) &= (1, x, 0), \\ Y_2(x, \xi) &= (-1/2, x/2, 0), \\ Y_i(x, \xi) &= (0, v_i(x, \xi), 0), \quad 3 \le i \le n+1, \\ Y_{n+2}(x, \xi) &= (0, \xi/2, -1/2) \\ Y_{n+3}(x, \xi) &= (0, \xi, 1). \end{aligned} \tag{4.21}$$

We want to determine certain of the Maurer–Cartan forms ω_a^b by computing dY_a on the basis given in (4.20). In particular, we compute the derivatives dY_1 and dY_{n+3} and find

$$dY_1(v_i, 0) = (0, v_i, 0) = Y_i,$$
$$dY_1(0, v_i) = (0, 0, 0),$$
$$dY_1(\xi, -x) = (0, \xi, 0) = Y_{n+2} + (1/2)Y_{n+3},$$

(4.22)

and

$$dY_{n+3}(v_i, 0) = (0, 0, 0),$$
$$dY_{n+3}(0, v_i) = (0, v_i, 0) = Y_i,$$
$$dY_{n+3}(\xi, -x) = (0, -x, 0) = (-1/2)Y_1 - Y_2.$$

(4.23)

Comparing these equations with the equation

$$dY_a = \sum \omega_a^b Y_b,$$

we see that the 1-forms,

$$\{\omega_1^i, \omega_{n+3}^i, \omega_1^{n+2}\}, \quad 3 \le i \le n+1,$$

(4.24)

form the dual basis to the basis given in (4.20) for the tangent space to $T_1 S^n$ at (x, ξ). Since $(\xi, -x)$ has length $\sqrt{2}$, we have

$$\omega_1^{n+2}(X, Z) = ((X, Z) \cdot (\xi, -x))/2 = (X \cdot \xi - Z \cdot x)/2,$$

for a tangent vector (X, Z) to $T_1 S^n$ at (x, ξ). Using equation (4.5),

$$X \cdot \xi + Z \cdot x = 0,$$

we see that

$$\omega_1^{n+2}(X, Z) = X \cdot \xi,$$

(4.25)

so ω_1^{n+2} is precisely the form ω in equation (4.6). We now want to show that ω_1^{n+2} satisfies condition (4.2). This is a straightforward calculation using the Maurer–Cartan equation (4.19) for $d\omega_1^{n+2}$ and the skew-symmetry relations (4.18). By equation (4.19), we have

$$d\omega_1^{n+2} = \sum \omega_1^c \wedge \omega_c^{n+2}.$$

The skew-symmetry relations (4.18) imply that $\omega_1^2 = 0$ and $\omega_{n+3}^{n+2} = 0$. Furthermore, in computing $(d\omega_1^{n+2})^{n-1}$, we can ignore any term involving ω_1^{n+2}, since we will eventually take the wedge product with ω_1^{n+2}. Thus in computing the wedge product $d\omega_1^{n+2} \wedge d\omega_1^{n+2}$, we need only to consider

$$\left(\sum d\omega_1^i \wedge d\omega_i^{n+2}\right) \wedge \left(\sum d\omega_1^j \wedge d\omega_j^{n+2}\right).$$

If $i \ne j$, we have a term of the form

$$\omega_1^i \wedge \omega_i^{n+2} \wedge \omega_1^j \wedge \omega_j^{n+2} = \omega_1^i \wedge (-\omega_{n+3}^i) \wedge \omega_1^j \wedge (-\omega_{n+3}^j) = \omega_1^i \wedge \omega_{n+3}^i \wedge \omega_1^j \wedge \omega_{n+3}^j \ne 0,$$

where the sign changes are due to the skew-symmetry relations (4.18). The last term is nonzero since each of the factors is in the basis given in (4.24). Thus we have

$$d\omega_1^{n+2} \wedge d\omega_1^{n+2} = 2 \sum_{i<j} \omega_1^i \wedge \omega_{n+3}^i \wedge \omega_1^j \wedge \omega_{n+3}^j \quad (\text{mod } \omega_1^{n+2}). \tag{4.26}$$

One continues this process by taking the wedge product of (4.26) with $d\omega_1^{n+2}$. This time there are three sign changes in each term as a result of the skew-symmetry relations (4.18), and we get

$$(d\omega_1^{n+2})^3 = (-1)^3 (3!) \sum_{i<j<k} \omega_1^i \wedge \omega_{n+3}^i \wedge \omega_1^j \wedge \omega_{n+3}^j \wedge \omega_1^k \wedge \omega_{n+3}^k \quad (\text{mod } \omega_1^{n+2}).$$

Continuing this process, one eventually obtains

$$\omega_1^{n+2} \wedge (d\omega_1^{n+2})^{n-1} = \omega_1^{n+2} \wedge \left(\sum \omega_1^i \wedge \omega_i^{n+2} \right)^{n-1} \tag{4.27}$$

$$= (-1)^{n-1}(n-1)! \quad \omega_1^{n+2} \wedge \omega_1^3 \wedge \omega_{n+3}^3 \wedge \cdots \wedge \omega_1^{n+1} \wedge \omega_{n+3}^{n+1} \neq 0.$$

The last form is nonzero because the set (4.24) is a basis for the cotangent space to $T_1 S^n$ at (x, ξ). Finally, note that the form

$$\omega_1^{n+2} = \langle dY_1, Y_{n+3} \rangle, \tag{4.28}$$

is globally defined on $T_1 S^n$, since Y_1 and Y_{n+3} are globally defined by equation (4.21), even though the rest of the Lie frame is only defined on the open set U.

As we noted above, we can use the diffeomorphism given in (4.8) to transfer this Lie frame and contact form ω_1^{n+2} to the manifold Λ^{2n-1} of lines on the Lie quadric. Now suppose that $\{Z_1, \ldots, Z_{n+3}\}$ is an arbitrary Lie frame on the open set U with the property that the line $[Z_1, Z_{n+3}]$ equals the line $[Y_1, Y_{n+3}]$ at all points of U, i.e.,

$$Z_1 = \alpha Y_1 + \beta Y_{n+3}, \qquad Z_{n+3} = \gamma Y_1 + \delta Y_{n+3}, \tag{4.29}$$

for smooth functions $\alpha, \beta, \gamma, \delta$ with $\alpha\delta - \beta\gamma \neq 0$ on U. Let $\{\theta_a^b\}$ be the Maurer–Cartan forms for this Lie frame. Then using the scalar product relations (4.11), we get

$$\theta_1^{n+2} = \langle dZ_1, Z_{n+3} \rangle = \langle d(\alpha Y_1 + \beta Y_{n+3}), \gamma Y_1 + \delta Y_{n+3} \rangle$$

$$= \alpha\delta \langle dY_1, Y_{n+3} \rangle + \beta\gamma \langle dY_{n+3}, Y_1 \rangle = \alpha\delta \omega_1^{n+2} + \beta\gamma \omega_{n+3}^2 \tag{4.30}$$

$$= (\alpha\delta - \beta\gamma)\omega_1^{n+2}.$$

Thus, θ_1^{n+2} is also a contact form on $T_1 S^n$.

4.2 Definition of Legendre Submanifolds

In the last section, we showed that $T_1 S^n$ (and hence Λ^{2n-1}) is a contact manifold. We begin this section by proving a basic result concerning contact manifolds in general

(Theorem 4.1). Let V^{2n-1} be a contact manifold with contact form ω. Let D be the corresponding contact distribution defined by

$$D_p = \{Y \in T_p V^{2n-1} \mid \omega(Y) = 0\},$$

for $p \in V^{2n-1}$. An immersion $\phi : W^k \to V^{2n-1}$ of a smooth k-dimensional manifold W^k into V^{2n-1} is called an *integral submanifold* of the distribution D if $\phi^*\omega = 0$ on W^k, i.e., for each tangent vector Y at each point $w \in W$, the vector $d\phi(Y)$ is in the distribution D at the point $\phi(w)$. (See Blair [9, p. 36].)

Theorem 4.1. *Let V^{2n-1} be a contact manifold with contact form ω. Then there exist integral submanifolds of the contact distribution D of dimension $n - 1$, but none of higher dimension.*

Proof. This is a local result, and the key tool in the local study of contact manifolds in Darboux's theorem (see, for example, Arnold [3, p. 362] or Sternberg [176, p. 141]), which states that for every point of a contact manifold V^{2n-1}, there is a local coordinate neighborhood U with coordinates (x_i, y_i, z), $1 \le i \le n - 1$, on which the contact form satisfies

$$\omega = dz - \sum_{i=1}^{n-1} y_i dx_i.$$

Now, given any point in U with coordinates (x_i^0, y_i^0, z^0), the slice defined by the equations

$$x_i = x_i^0, \quad z = z^0, \quad 1 \le i \le n - 1,$$

clearly defines an $(n-1)$-dimensional integral submanifold of D containing the given point.

Conversely, suppose that W^k is an immersed k-dimensional integral submanifold of D with $k > n - 1$. Since this is a local result, we may consider W^k to be embedded in V^{2n-1}. Let $X_1, \ldots, X_k$ be linearly independent vector fields tangent to W^k on some open set Ω in W^k. Let $X_{k+1}, \ldots, X_{2n-1}$ be tangent vectors to V^{2n-1} at a point $w \in \Omega$ such that $\{X_1, \ldots, X_{2n-1}\}$ is a basis for $T_w V^{2n-1}$. Since the tangent distribution to W^k is integrable, the Lie bracket $[X_i, X_j]$ is also tangent to W^k, for $1 \le i, j \le k$. Furthermore, since the tangent distribution to W^k is contained in the distribution D, we have

$$\omega(X_i) = 0, \quad \omega([X_i, X_j]) = 0,$$

and thus,

$$d\omega(X_i, X_j) = X_i \omega(X_j) - X_j \omega(X_i) - \omega([X_i, X_j]) = 0,$$

for $1 \le i, j \le k$, on Ω. Since $k > n - 1$, this implies that at the point w, we have

$$\omega \wedge (d\omega)^{n-1}(X_1, \ldots, X_{2n-1}) = 0,$$

contradicting the assumption that ω is a contact form on V^{2n-1}. □

An immersed $(n-1)$-dimensional integral submanifold of the contact distribution D is called a *Legendre submanifold*. We now return to our specific case of the contact manifold $T_1 S^n$. We first want to formulate necessary and sufficient conditions for a smooth map $\mu : M^{n-1} \to T_1 S^n$ to be a Legendre submanifold. We consider $T_1 S^n$ as a submanifold of $S^n \times S^n$ as in equation (4.1). Thus we can write $\mu = (f, \xi)$, where f and ξ are both smooth maps from M^{n-1} to S^n.

Theorem 4.2. *A smooth map $\mu = (f, \xi)$ from an $(n-1)$-dimensional manifold M^{n-1} into $T_1 S^n$ is a Legendre submanifold if and only if the following three conditions are satisfied.*

(1) Scalar product conditions: $f \cdot f = 1, \quad \xi \cdot \xi = 1, \quad f \cdot \xi = 0.$
(2) Immersion condition: *There is no nonzero tangent vector X at any point $x \in M^{n-1}$ such that $df(X)$ and $d\xi(X)$ are both equal to zero.*
(3) Contact condition: $df \cdot \xi = 0.$

Proof. By equation (4.1), the scalar product conditions are precisely the conditions necessary for the image of the map $\mu = (f, \xi)$ to be contained in $T_1 S^n$. Next, since

$$d\mu(X) = (df(X), d\xi(X)),$$

the second condition is precisely what is needed for μ to be an immersion. Finally, from equation (4.6) we have

$$\omega(d\mu(X)) = df(X) \cdot \xi(x),$$

for each $X \in T_x M^{n-1}$. Hence the condition $\mu^* \omega = 0$ on M^{n-1} is equivalent to the third condition above. $\square$

We now want to translate these conditions into the projective setting, and find necessary and sufficient conditions for a smooth map $\lambda : M^{n-1} \to \Lambda^{2n-1}$ to be a Legendre submanifold. We again make use of the diffeomorphism defined in equation (4.8) between $T_1 S^n$ and Λ^{2n-1}. For each $x \in M^{n-1}$, we know that $\lambda(x)$ is a line on the quadric Q^{n+1}. This line contains exactly one point $[Y_1(x)]$ corresponding to a point sphere in S^n and one point $[Y_{n+3}(x)]$ corresponding to a great sphere in S^n. The map $[Y_1]$ from M^{n-1} to Q^{n+1} is called the *Möbius projection* or *point sphere map* of λ, and likewise, the map $[Y_{n+3}]$ is called the *great sphere map*.

The homogeneous coordinates of these points with respect to the standard basis are given by

$$Y_1(x) = (1, f(x), 0), \qquad Y_{n+3}(x) = (0, \xi(x), 1), \qquad (4.31)$$

where f and ξ are both smooth maps from M^{n-1} to S^n defined by formula (4.31). The map f is called the *spherical projection* of λ, and ξ is called the *spherical field of unit normals*. The maps f and ξ depend on the choice of orthonormal basis $\{e_1, \ldots, e_{n+2}\}$ for the orthogonal complement of e_{n+3}. In this way, λ determines a map $\mu = (f, \xi)$ from M^{n-1} to $T_1 S^n$, and because of the diffeomorphism (4.8), λ is a Legendre submanifold if and only if μ satisfies the conditions of Theorem 4.2. However, it is useful to have conditions for when λ determines a Legendre submanifold that do not depend on the special parametrization of λ by $[Y_1, Y_{n+3}]$. In fact, in most applications of Lie sphere geometry to submanifolds of S^n or $\mathbf{R}^n$, it is better to use a Lie frame $\{Z_1, \ldots, Z_{n+3}\}$ with $\lambda = [Z_1, Z_{n+3}]$, where Z_1 and Z_{n+3} are not the point sphere and great sphere maps. The following projective formulation of the conditions needed for a Legendre submanifold was given by Pinkall [150], where he referred to a Legendre submanifold as a "Lie geometric hypersurface." The three conditions correspond, respectively, to the three conditions in Theorem 4.2.

Theorem 4.3. *Let* $\lambda : M^{n-1} \to \Lambda^{2n-1}$ *be a smooth map with* $\lambda = [Z_1, Z_{n+3}]$, *where* Z_1 *and* Z_{n+3} *are smooth maps from* M^{n-1} *into* $\mathbf{R}_2^{n+3}$. *Then* λ *determines a Legendre submanifold if and only if* Z_1 *and* Z_{n+3} *satisfy the following conditions:*

(1) Scalar product conditions: *For each* $x \in M^{n-1}$, *the vectors* $Z_1(x)$ *and* $Z_{n+3}(x)$ *are linearly independent and*

$$\langle Z_1, Z_1 \rangle = 0, \quad \langle Z_{n+3}, Z_{n+3} \rangle = 0, \quad \langle Z_1, Z_{n+3} \rangle = 0.$$

(2) Immersion condition: *There is no nonzero tangent vector* X *at any point* $x \in M^{n-1}$ *such that* $dZ_1(X)$ *and* $dZ_{n+3}(X)$ *are both in* $\mathrm{Span}\{Z_1(x), Z_{n+3}(x)\}$.
(3) Contact condition: $\langle dZ_1, Z_{n+3} \rangle = 0$.

These conditions are invariant under a reparametrization $\lambda = [W_1, W_{n+3}]$, *where* $W_1 = \alpha Z_1 + \beta Z_{n+3}$ *and* $W_{n+3} = \gamma Z_1 + \delta Z_{n+3}$, *for smooth functions* $\alpha, \beta, \gamma, \delta$ *on* M^{n-1} *with* $\alpha\delta - \beta\gamma \neq 0$.

Proof. The proof is accomplished in two steps. First, we know that the map λ induces two maps Y_1 and Y_{n+3} as equation (4.31), which in turn determine f and ξ. We show that the pair $\{Y_1, Y_{n+3}\}$ satisfies conditions (1)–(3) above if and only if the map $\mu = (f, \xi)$ satisfies the conditions in Theorem 4.2. Secondly, we show that the conditions (1)–(3) above are invariant under a transformation to $\{W_1, W_{n+3}\}$ as above. In particular, the conditions are satisfied by an arbitrary pair $\{Z_1, Z_{n+3}\}$ if and only if they are satisfied by $\{Y_1, Y_{n+3}\}$.
     First, consider $Y_1 = (1, f, 0)$, $Y_{n+3} = (0, \xi, 1)$. Then,

$$\langle Y_1, Y_1 \rangle = |f|^2 - 1, \quad \langle Y_{n+3}, Y_{n+3} \rangle = |\xi|^2 - 1, \quad \langle Y_1, Y_{n+3} \rangle = f \cdot \xi.$$

Thus, condition (1) for the pair $\{Y_1, Y_{n+3}\}$ is equivalent to the scalar product condition for the pair (f, ξ) in Theorem 4.2. Next, suppose that X is a nonzero vector in $T_x M^{n-1}$. Then

$$dY_1(X) = (0, df(X), 0), \qquad dY_{n+3}(X) = (0, d\xi(X), 0). \qquad (4.32)$$

Hence, $dY_1(X)$ and $dY_{n+3}(X)$ are in $\mathrm{Span}\{Y_1(x), Y_{n+3}(x)\}$ if and only if they are zero, i.e., $df(X) = 0$ and $d\xi(X) = 0$. Therefore, the pair $\{Y_1, Y_{n+3}\}$ satisfies condition (2) if and only if the pair (f, ξ) satisfies the immersion condition of Theorem 4.2. Finally, from equation (4.32), we have

$$\langle dY_1(X), Y_{n+3}(x) \rangle = df(X) \cdot \xi(x),$$

so the pair $\{Y_1, Y_{n+3}\}$ satisfies condition (3) if and only if (f, ξ) satisfies the contact condition of Theorem 4.2.

Now suppose that a pair $\{Z_1, Z_{n+3}\}$ satisfies conditions (1)–(3) and that $\{W_1, W_{n+3}\}$ is given as in the statement of the theorem. It follows from Theorem 2.6 of Chapter 2, p. 22, that condition (1) is precisely the condition that $[Z_1(x)]$ and $[Z_{n+3}(x)]$ be two distinct points on the quadric Q^{n+1} such that the line $[Z_1(x), Z_{n+3}(x)]$ lies on Q^{n+1}. This clearly holds for the pair $\{W_1, W_{n+3}\}$ if and only if it holds for $\{Z_1, Z_{n+3}\}$, since $[W_1, W_{n+3}] = [Z_1, Z_{n+3}]$. Next, we compute

$$dW_1 = \alpha dZ_1 + \beta dZ_{n+3} + (d\alpha)Z_1 + (d\beta)Z_{n+3}, \qquad (4.33)$$
$$dW_{n+3} = \gamma dZ_1 + \delta dZ_{n+3} + (d\gamma)Z_1 + (d\delta)Z_{n+3}.$$

The condition $\alpha\delta - \beta\gamma \neq 0$ allows us to solve for dZ_1 and dZ_{n+3} in terms of dW_1 and dW_{n+3}, mod $\{Z_1, Z_{n+3}\}$. From this, it is clear that condition (2) holds for the pair $\{W_1, W_{n+3}\}$ if and only if it holds for the pair $\{Z_1, Z_{n+3}\}$. Finally, note that the scalar product condition $\langle Z_1, Z_{n+3} \rangle = 0$ implies that

$$\langle dZ_{n+3}, Z_1 \rangle = -\langle dZ_1, Z_{n+3} \rangle.$$

Using this and the scalar product relations (1), we compute

$$
\begin{aligned}
\langle dW_1, W_{n+3} \rangle &= \langle \alpha dZ_1 + \beta dZ_{n+3} + (d\alpha)Z_1 + (d\beta)Z_{n+3}, \gamma dZ_1 + \delta dZ_{n+3} \rangle \\
&= \alpha\delta\langle dZ_1, Z_{n+3} \rangle + \beta\gamma\langle dZ_{n+3}, Z_1 \rangle \qquad (4.34) \\
&= (\alpha\delta - \beta\gamma)\langle dZ_1, Z_{n+3} \rangle.
\end{aligned}
$$

Thus, $\{W_1, W_{n+3}\}$ satisfies condition (3) precisely when $\{Z_1, Z_{n+3}\}$ satisfies condition (3). □

4.3 The Legendre Map

All oriented hypersurfaces in the sphere S^n, Euclidean space $\mathbf{R}^n$ or hyperbolic space H^n naturally induce Legendre submanifolds of Λ^{2n-1}, as do all submanifolds of codimension $m > 1$ in these spaces. In this section, we study these examples and see, conversely, how a Legendre submanifold naturally induces a smooth map into S^n which may have singularities.

First, suppose that $f : M^{n-1} \to S^n$ is an immersed oriented hypersurface with field of unit normals $\xi : M^{n-1} \to S^n$. The induced Legendre submanifold is given by the map $\lambda : M^{n-1} \to \Lambda^{2n-1}$ defined by

$$\lambda(x) = [Y_1(x), Y_{n+3}(x)],$$

where

$$Y_1(x) = (1, f(x), 0), \qquad Y_{n+3}(x) = (0, \xi(x), 1). \qquad (4.35)$$

The map λ is called the *Legendre map* induced by the immersion f with field of unit normals ξ. We will also refer to λ as the *Legendre submanifold induced by the pair* (f, ξ). It is easy to check that the pair $\{Y_1, Y_{n+3}\}$ satisfies the conditions of Theorem 4.3. Condition (1) is immediate since both f and ξ are maps into S^n, and $\xi(x)$ is tangent to S^n at $f(x)$ for each x in M^{n-1}. Condition (2) is satisfied since

$$dY_1(X) = (0, df(X), 0),$$

for any vector $X \in T_x M^{n-1}$. Since f is an immersion, $df(X) \neq 0$ for a nonzero vector X, and thus $dY_1(X)$ is not in $\text{Span}\{Y_1(x), Y_{n+3}(x)\}$. Finally, condition (3) is satisfied since

$$\langle dY_1(X), Y_{n+3}(x) \rangle = df(X) \cdot \xi(x) = 0,$$

because ξ is a field of unit normals to f.

Next, we handle the case of a submanifold $\phi : V \to S^n$ of codimension $m + 1$ greater than one. Let B^{n-1} be the unit normal bundle of the submanifold ϕ. Then B^{n-1} can be considered to be the submanifold of $V \times S^n$ given by

$$B^{n-1} = \{(x, \xi) | \phi(x) \cdot \xi = 0, \quad d\phi(X) \cdot \xi = 0, \quad \text{for all } X \in T_x V\}.$$

The *Legendre submanifold induced by* $\phi(V)$ is the map $\lambda : B^{n-1} \to \Lambda^{2n-1}$ defined by

$$\lambda(x, \xi) = [Y_1(x, \xi), Y_{n+3}(x, \xi)], \tag{4.36}$$

where

$$Y_1(x, \xi) = (1, \phi(x), 0), \qquad Y_{n+3}(x, \xi) = (0, \xi, 1). \tag{4.37}$$

Geometrically, $\lambda(x, \xi)$ is the line on the quadric Q^{n+1} corresponding to the parabolic pencil of spheres in S^n in oriented contact at the contact element $(\phi(x), \xi) \in T_1 S^n$. As in the case of a hypersurface, condition (1) is easily checked. However, condition (2) is somewhat different. To compute the differentials of Y_1 and Y_{n+3} at a given point (x, ξ), we first construct a local trivialization of B^{n-1} in a neighborhood of (x, ξ). Let $\{\xi_0, \ldots, \xi_m\}$ be an orthonormal frame at x with $\xi_0 = \xi$. Let W be a normal coordinate neighborhood of x in V, as defined in Kobayashi–Nomizu [95, Vol. 1, p. 148], and extend $\xi_0, \ldots, \xi_m$ to orthonormal normal vector fields on W by parallel translation with respect to the normal connection along geodesics in V through x. For any point $w \in W$ and unit normal η to $\phi(V)$ at w, we can write

$$\eta = (1 - \sum_{i=1}^{m} t_i^2)^{1/2} \xi_0 + t_1 \xi_1 + \cdots + t_m \xi_m,$$

where $0 \leq |t_i| \leq 1$, for all i, and $t_1^2 + \cdots + t_m^2 \leq 1$. The tangent space to B^{n-1} at the given point (x, ξ) can be considered to be

$$T_x V \times \text{Span}\{\partial/\partial t_1, \ldots, \partial/\partial t_m\} = T_x \times \mathbf{R}^m. \tag{4.38}$$

Since $\xi_0(x) = \xi$, and ξ_0 is parallel with respect to the normal connection, we have for $X \in T_x V$,

$$d\xi_0(X) = d\phi(-A_\xi X),$$

where A_ξ is the shape operator determined by ξ. Thus, we have

$$dY_1(X, 0) = (0, d\phi(X), 0), \tag{4.39}$$
$$dY_{n+3}(X, 0) = (0, d\xi_0(X), 0) = (0, d\phi(-A_\xi X), 0).$$

Next we compute from equation (4.37),

$$dY_1(0, Z) = (0, 0, 0), \qquad dY_{n+3}(0, Z) = (0, Z, 0). \tag{4.40}$$

From equations (4.39) and (4.40), we see that there is no nonzero vector (X, Z) such that $dY_1(X, Z)$ and $dY_{n+3}(X, Z)$ are both in Span$\{Y_1, Y_{n+3}\}$, and so condition (2) is satisfied. Finally, condition (3) holds since

$$\langle dY_1(X, Z), Y_{n+3}(x, \xi) \rangle = d\phi(X) \cdot \xi = 0.$$

The situation for submanifolds of $\mathbf{R}^n$ or H^n is similar. First, suppose that $F : M^{n-1} \to \mathbf{R}^n$ is an oriented hypersurface with field of unit normals $\eta : M^{n-1} \to \mathbf{R}^n$. As usual, we identify $\mathbf{R}^n$ with the subspace of $\mathbf{R}_2^{n+3}$ spanned by $\{e_3, \ldots, e_{n+2}\}$. The Legendre submanifold induced by (F, η) is the map $\lambda : M^{n-1} \to \Lambda^{2n-1}$ defined by $\lambda = [Y_1, Y_{n+3}]$, where

$$Y_1 = (1 + F \cdot F, 1 - F \cdot F, 2F, 0)/2, \qquad Y_{n+3} = (F \cdot \eta, -(F \cdot \eta), \eta, 1). \tag{4.41}$$

By equation (2.14) of Chapter 2, p. 16, $[Y_1(x)]$ corresponds to the point sphere and $[Y_{n+3}(x)]$ corresponds to the hyperplane in the parabolic pencil determined by the line $\lambda(x)$ for each $x \in M^{n-1}$. The reader can easily verify conditions (1)–(3) of Theorem 4.3 in a manner similar to the spherical case. In the case of a submanifold $\psi : V \to \mathbf{R}^n$ of codimension greater than one, the induced Legendre submanifold is the map λ from the unit normal bundle B^{n-1} to Λ^{2n-1} defined by

$$\lambda(x, \eta) = [Y_1(x, \eta), Y_{n+3}(x, \eta)],$$

where

$$Y_1(x, \eta) = (1 + \psi(x) \cdot \psi(x), 1 - \psi(x) \cdot \psi(x), 2\psi(x), 0)/2, \tag{4.42}$$
$$Y_{n+3}(x, \eta) = (\psi(x) \cdot \eta, -(\psi(x) \cdot \eta), \eta, 1).$$

The verification that the pair $\{Y_1, Y_{n+3}\}$ satisfies conditions (1)–(3) is similar to that for submanifolds of S^n of codimension greater than one.

Finally, as in Section 2.4, we consider H^n to be the submanifold of the Lorentz space $\mathbf{R}_1^{n+1}$ spanned by $\{e_1, e_3, \ldots, e_{n+2}\}$ defined as follows:

$$H^n = \{y \in \mathbf{R}_1^{n+1} | (y, y) = -1, \ y_1 \geq 1\},$$

where $(,)$ is the Lorentz metric on $\mathbf{R}_1^{n+1}$ obtained by restricting the Lie metric. Let $h : M^{n-1} \to H^n$ be an oriented hypersurface with field of unit normals $\zeta : M^{n-1} \to \mathbf{R}_1^{n+1}$. The Legendre submanifold induced by (h, ζ) is given by the map

$$\lambda : M^{n-1} \to \Lambda^{2n-1},$$

defined by $\lambda = [Y_1, Y_{n+3}]$, where

$$Y_1(x) = h(x) + e_2, \qquad Y_{n+3}(x) = \zeta(x) + e_{n+3}. \tag{4.43}$$

Note that $(h, h) = -1$, so $\langle Y_1, Y_1 \rangle = 0$, while $(\zeta, \zeta) = 1$, so $\langle Y_{n+3}, Y_{n+3} \rangle = 0$. The reader can easily check that the conditions (1)–(3) are satisfied. Finally, if $\gamma : V \to H^n$ is an immersed submanifold of codimension greater than one, then the Legendre submanifold $\lambda : B^{n-1} \to \Lambda^{2n-1}$ is again defined on the unit normal bundle B^{n-1} of the submanifold $\gamma(V)$ in the obvious way.

Now suppose that $\lambda : M^{n-1} \to \Lambda^{2n-1}$ is an arbitrary Legendre submanifold. As we have seen, it is always possible to parametrize λ by the point sphere map $[Y_1]$ and the great sphere map $[Y_{n+3}]$ given by

$$Y_1 = (1, f, 0), \qquad Y_{n+3} = (0, \xi, 1). \tag{4.44}$$

This defines two maps f and ξ from M^{n-1} to S^n, which we called the spherical projection and spherical field of unit normals, respectively, in Section 4.2. Both f and ξ are smooth maps, but neither need be an immersion or even have constant rank. (See Example 4.4 below.) The Legendre submanifold induced by an oriented hypersurface in S^n is the special case where the spherical projection f is an immersion, i.e., f has constant rank $n - 1$ on M^{n-1}. In the case of the Legendre submanifold induced by a submanifold $\phi : V^k \to S^n$, the spherical projection $f : B^{n-1} \to S^n$ defined by $f(x, \xi) = \phi(x)$ has constant rank k.

If the range of the point sphere map $[Y_1]$ does not contain the improper point $[(1, -1, 0, \ldots, 0)]$, then λ also determines a *Euclidean projection*,

$$F : M^{n-1} \to \mathbf{R}^n,$$

and a *Euclidean field of unit normals*,

$$\eta : M^{n-1} \to \mathbf{R}^n.$$

These are defined by the equation $\lambda = [Z_1, Z_{n+3}]$, where

$$Z_1 = (1 + F \cdot F, 1 - F \cdot F, 2F, 0)/2, \qquad Z_{n+3} = (F \cdot \eta, -(F \cdot \eta), \eta, 1). \tag{4.45}$$

Here $[Z_1(x)]$ corresponds to the unique point sphere in the parabolic pencil determined by $\lambda(x)$, and $[Z_{n+3}(x)]$ corresponds to the unique plane in this pencil. As in the spherical case, the smooth maps F and η need not have constant rank. Finally, if the range of the Euclidean projection F lies inside some disk Ω in $\mathbf{R}^n$, then one can define a hyperbolic projection and hyperbolic field of unit normals by placing a hyperbolic metric on Ω.

Example 4.4 (*a Euclidean projection that is not an immersion*). An example where the Euclidean (or spherical) projection does not have constant rank is illustrated by the cyclide of Dupin in Figure 4.1. Here the corresponding Legendre submanifold is a map $\lambda : T^2 \rightarrow \Lambda^5$, where T^2 is a two-dimensional torus. The Euclidean projection $F : T^2 \rightarrow \mathbf{R}^3$ maps the circle S^1 containing the points A, B, C and D to the point P. However, the map λ into the space of lines on the quadric (corresponding to contact elements) is an immersion. The four arrows in Figure 4.1 represent the contact elements corresponding under the map λ to the four points indicated on the circle S^1. Actually, examples of Legendre submanifolds whose Euclidean or spherical projection is not an immersion are plentiful, as will be seen in the next section.

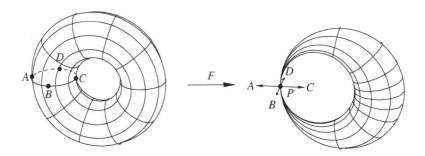

Fig. 4.1. A Euclidean projection F with a singularity.

4.4 Curvature Spheres and Parallel Submanifolds

To motivate the definition of a curvature sphere, we consider the case of an oriented hypersurface $f : M^{n-1} \rightarrow S^n$ with field of unit normals $\xi : M^{n-1} \rightarrow S^n$. The shape operator of f at a point $x \in M^{n-1}$ is the symmetric linear transformation $A : T_x M^{n-1} \rightarrow T_x M^{n-1}$ defined by the equation

$$df(AX) = -d\xi(X), \quad X \in T_x M^{n-1}. \tag{4.46}$$

The eigenvalues of A are called the *principal curvatures*, and the corresponding eigenvectors are called the *principal vectors*. We next recall the notion of a focal point of an immersion. For each real number t, define a map

$$f_t : M^{n-1} \rightarrow S^n,$$

by

$$f_t = \cos t \, f + \sin t \, \xi. \tag{4.47}$$

For each $x \in M^{n-1}$, the point $f_t(x)$ lies an oriented distance t along the normal geodesic to $f(M^{n-1})$ at $f(x)$. A point $p = f_t(x)$ is called a *focal point of multiplicity* $m > 0$ of f at x if the nullity of df_t is equal to m at x. Geometrically, one thinks of focal points as points where nearby normal geodesics intersect. It is well known that the location of focal points is related to the principal curvatures. Specifically, if $X \in T_x M^{n-1}$, then by equation (4.46) we have

$$df_t(X) = \cos t \, df(X) + \sin t \, d\xi(X) = df(\cos t \, X - \sin t \, AX). \qquad (4.48)$$

Thus, $df_t(X)$ equals zero for $X \neq 0$ if and only if $\cot t$ is a principal curvature of f at x, and X is a corresponding principal vector. Hence, $p = f_t(x)$ is a focal point of f at x of multiplicity m if and only if $\cot t$ is a principal curvature of multiplicity m at x. Note that each principal curvature

$$\kappa = \cot t, \quad 0 < t < \pi,$$

produces two distinct antipodal focal points on the normal geodesic with parameter values t and $t + \pi$. The oriented hypersphere centered at a focal point p and in oriented contact with $f(M^{n-1})$ at $f(x)$ is called a *curvature sphere* of f at x. The two antipodal focal points determined by κ are the two centers of the corresponding curvature sphere. Thus, the correspondence between principal curvatures and curvature spheres is bijective. The multiplicity of the curvature sphere is by definition equal to the multiplicity of the corresponding principal curvature.

We now consider these ideas as they apply to the Legendre submanifold induced by the oriented hypersurface determined by f and ξ. As in equation (4.35), we have $\lambda = [Y_1, Y_{n+3}]$, where

$$Y_1 = (1, f, 0), \qquad Y_{n+3} = (0, \xi, 1). \qquad (4.49)$$

For each $x \in M^{n-1}$, the points on the line $\lambda(x)$ can be parametrized as

$$[K_t(x)] = [\cos t \, Y_1(x) + \sin t \, Y_{n+3}(x)] = [(\cos t, \, f_t(x), \, \sin t)], \qquad (4.50)$$

where f_t is given in equation (4.47). By equation (2.21) of Chapter 2, p. 17, the point $[K_t(x)]$ in Q^{n+1} corresponds to the oriented sphere in S^n with center $f_t(x)$ and signed radius t. This sphere is in oriented contact with the oriented hypersurface $f(M^{n-1})$ at $f(x)$. Given a tangent vector $X \in T_x M^{n-1}$, we have

$$dK_t(X) = (0, df_t(X), 0). \qquad (4.51)$$

Thus, $dK_t(X) = (0, 0, 0)$ if and only if $df_t(X) = 0$, i.e., $p = f_t(x)$ is a focal point of f at x. Hence, we have shown the following.

Lemma 4.5. *The point $[K_t(x)]$ in Q^{n+1} corresponds to a curvature sphere of the hypersurface f at x if and only if $dK_t(X) = (0, 0, 0)$ for some nonzero vector $X \in T_x M^{n-1}$.*

This characterization of curvature spheres depends on the parametrization of λ given by $\{Y_1, Y_{n+3}\}$ as in equation (4.49), and it has only been defined in the case where the spherical projection f is an immersion. Since it is often desirable to use a different parametrization of λ, we would like a definition of curvature sphere which is independent of the parametrization of λ. We would also like a definition that is valid for an arbitrary Legendre submanifold. This definition is given in the following paragraph.

Let $\lambda : M^{n-1} \to \Lambda^{2n-1}$ be a Legendre submanifold parametrized by the pair $\{Z_1, Z_{n+3}\}$, as in Theorem 4.3. Let $x \in M^{n-1}$ and $r, s \in \mathbf{R}$ with $(r, s) \neq (0, 0)$. The sphere,

$$[K] = [rZ_1(x) + sZ_{n+3}(x)],$$

is called a *curvature sphere* of λ at x if there exists a nonzero vector X in $T_x M^{n-1}$ such that

$$r \, dZ_1(X) + s \, dZ_{n+3}(X) \in \text{Span}\{Z_1(x), Z_{n+3}(x)\}. \qquad (4.52)$$

The vector X is called a *principal vector* corresponding to the curvature sphere $[K]$. By equation (4.33), this definition is invariant under a change of parametrization of the form considered in Theorem 4.3. Furthermore, if we take the special parametrization $Z_1 = Y_1$, $Z_{n+3} = Y_{n+3}$ given in equation (4.49), then condition (4.52) holds if and only if $r \, dY_1(X) + s \, dY_{n+3}(X)$ actually equals $(0, 0, 0)$. Thus, this definition is a generalization of the condition in Lemma 4.5.

From equation (4.52), it is clear that the set of principal vectors corresponding to a given curvature sphere $[K]$ at x is a subspace of $T_x M^{n-1}$. This set is called the *principal space* corresponding to the curvature sphere $[K]$. Its dimension is the *multiplicity* of $[K]$.

Remark 4.6. The definition of curvature sphere can be developed in the context of Lie sphere geometry without any reference to submanifolds of S^n (see Cecil–Chern [37] for details). In that case, one begins with a Legendre submanifold $\lambda : M^{n-1} \to \Lambda^{2n-1}$ and considers a curve $\gamma(t)$ lying in M^{n-1}. The set of points in Q^{n+1} lying on the set of lines $\lambda(\gamma(t))$ forms a ruled surface in Q^{n+1}. One then considers conditions for this ruled surface to be developable. This leads to a system of linear equations whose roots determine the curvature spheres at each point along the curve.

We next want to show that the notion of curvature sphere is invariant under Lie sphere transformations. Let $\lambda : M^{n-1} \to \Lambda^{2n-1}$ be a Legendre submanifold parametrized by $\lambda = [Z_1, Z_{n+3}]$. Suppose $\beta = P(B)$ is the Lie sphere transformation induced by an orthogonal transformation B in the group $O(n + 1, 2)$. Since B is orthogonal, it is easy to check that the maps, $W_1 = BZ_1$, $W_{n+3} = BZ_{n+3}$, satisfy the conditions (1)–(3) of Theorem 4.3. We will denote the Legendre submanifold defined by $\{W_1, W_{n+3}\}$ by

$$\beta\lambda : M^{n-1} \to \Lambda^{2n-1}.$$

The Legendre submanifolds λ and $\beta\lambda$ are said to be *Lie equivalent*. In terms of Euclidean geometry, suppose that V and W are two immersed submanifolds of $\mathbf{R}^n$ (or of S^n or H^n). We say that V and W are *Lie equivalent* if their induced Legendre submanifolds are Lie equivalent.

Consider λ and β as above, so that $\lambda = [Z_1, Z_{n+3}]$ and $\beta\lambda = [W_1, W_{n+3}]$. Note that for a tangent vector $X \in T_x M^{n-1}$ and for real numbers $(r, s) \neq (0, 0)$, we have

$$r \, dW_1(X) + s \, dW_{n+3}(X) = B(r \, dZ_1(X) + s \, dZ_{n+3}(X)), \qquad (4.53)$$

since B is linear. Thus, we see that

$$r\, dW_1(X) + s\, dW_{n+3}(X) \in \mathrm{Span}\{W_1(x), W_{n+3}(x)\}$$

if and only if

$$r\, dZ_1(X) + s\, dZ_{n+3}(X) \in \mathrm{Span}\{Z_1(x), Z_{n+3}(x)\}.$$

This immediately implies the following theorem.

Theorem 4.7. *Let* $\lambda : M^{n-1} \to \Lambda^{2n-1}$ *be a Legendre submanifold and* β *a Lie sphere transformation. The point* $[K]$ *on the line* $\lambda(x)$ *is a curvature sphere of* λ *at* x *if and only if the point* $\beta[K]$ *is a curvature sphere of the Legendre submanifold* $\beta\lambda$ *at* x. *Furthermore, the principal spaces corresponding to* $[K]$ *and* $\beta[K]$ *are identical.*

An important special case is when the Lie sphere transformation is a spherical parallel transformation P_t, as given in Section 3.5,

$$\begin{aligned}
P_t e_1 &= \cos t\, e_1 + \sin t\, e_{n+3}, \\
P_t e_{n+3} &= -\sin t\, e_1 + \cos t\, e_{n+3}, \\
P_t e_i &= e_i, \quad 2 \le i \le n+2.
\end{aligned} \tag{4.54}$$

Recall that P_t has the effect of adding t to the signed radius of each sphere in S^n while keeping the center fixed.

Suppose that $\lambda : M^{n-1} \to \Lambda^{2n-1}$ is a Legendre submanifold parametrized by the point sphere and great sphere maps $\{Y_1, Y_{n+3}\}$, as in equation (4.49). Then $P_t \lambda = [W_1, W_{n+3}]$, where

$$W_1 = P_t Y_1 = (\cos t, f, \sin t), \quad W_{n+3} = P_t Y_{n+3} = (-\sin t, \xi, \cos t). \tag{4.55}$$

Note that W_1 and W_{n+3} are not the point sphere and great sphere maps for $P_t\lambda$. Solving for the point sphere map Z_1 and the great sphere map Z_{n+3} of $P_t\lambda$, we find

$$\begin{aligned}
Z_1 &= \cos t\, W_1 - \sin t\, W_{n+3} = (1, \cos t\, f - \sin t\, \xi, 0), \\
Z_{n+3} &= \sin t\, W_1 + \cos t\, W_{n+3} = (0, \sin t\, f + \cos t\, \xi, 1).
\end{aligned} \tag{4.56}$$

From this, we see that $P_t\lambda$ has spherical projection and spherical unit normal field given, respectively, by

$$\begin{aligned}
f_{-t} &= \cos t\, f - \sin t\, \xi = \cos(-t)f + \sin(-t)\xi, \\
\xi_{-t} &= \sin t\, f + \cos t\, \xi = -\sin(-t)f + \cos(-t)\xi.
\end{aligned} \tag{4.57}$$

The minus sign occurs because P_t takes a sphere with center $f_{-t}(x)$ and radius $-t$ to the point sphere $f_{-t}(x)$. We call $P_t\lambda$ a *parallel submanifold* of λ. Formula (4.57) shows the close correspondence between these parallel submanifolds and the *parallel hypersurfaces* f_t to f, in the case where f is an immersed hypersurface. The spherical projection f_t has singularities at the focal points of f, but the parallel submanifold $P_t\lambda$ is still a smooth submanifold of Λ^{2n-1}. The following theorem, due to Pinkall [150, p. 428], shows that the number of these singularities is bounded for each $x \in M^{n-1}$.

Theorem 4.8. *Let* $\lambda : M^{n-1} \to \Lambda^{2n-1}$ *be a Legendre submanifold with spherical projection* f *and spherical unit normal field* ξ. *Then for each* $x \in M^{n-1}$, *the parallel map,*

$$f_t = \cos t \ f + \sin t \ \xi,$$

fails to be an immersion at x *for at most* $n - 1$ *values of* $t \in [0, \pi)$.

Here $[0, \pi)$ is the appropriate interval because of the phenomenon mentioned earlier that each principal curvature of an immersion produces two distinct antipodal focal points in the interval $[0, 2\pi)$. Before proving Pinkall's theorem, we state some important consequences which are obtained by passing to a parallel submanifold, if necessary, and then applying well-known results concerning immersed hypersurfaces in S^n.

Corollary 4.9. *Let* $\lambda : M^{n-1} \to \Lambda^{2n-1}$ *be a Legendre submanifold. Then*

(a) *at each point* $x \in M^{n-1}$, *there are at most* $n - 1$ *distinct curvature spheres* $K_1, \ldots, K_g$,
(b) *the principal vectors corresponding to a curvature sphere* K_i *form a subspace* T_i *of the tangent space* $T_x M^{n-1}$,
(c) *the tangent space* $T_x M^{n-1} = T_1 \oplus \cdots \oplus T_g$,
(d) *if the dimension of a given* T_i *is constant on an open subset* U *of* M^{n-1}, *then the principal distribution* T_i *is integrable on* U,
(e) *if* $\dim T_i = m > 1$ *on an open subset* U *of* M^{n-1}, *then the curvature sphere map* K_i *is constant along the leaves of the principal foliation* T_i.

Proof. In the case where the spherical projection f of λ is an immersion, the corollary follows from known results concerning hypersurfaces in S^n and the correspondence between the curvature spheres of λ and the principal curvatures of f. Specifically, (a)–(c) follow from elementary linear algebra applied to the (symmetric) shape operator A of the immersion f. As to (d) and (e), Ryan [163, p. 371] showed that the principal curvature functions on an immersed hypersurface are continuous. Nomizu [134] then showed that any continuous principal curvature function κ_i which has constant multiplicity on an open subset U in M^{n-1} is smooth, as is its corresponding principal distribution (see also, Singley [170]). If the multiplicity m_i of κ_i equals one on U, then T_i is integrable by the theory of ordinary differential equations. If $m_i > 1$, then the integrability of T_i, and the fact that κ_i is constant along the leaves of T_i are consequences of Codazzi's equation (Ryan [163], see also Cecil–Ryan [52, p. 139], and Reckziegel [157]–[159]).

Note that (a)–(c) are pointwise statements, while (d)–(e) hold on an open set U if they can be shown to hold in a neighborhood of each point of U. Now let x be an arbitrary point of M^{n-1}. If the spherical projection f is not an immersion at x, then by Theorem 4.8, we can find a parallel transformation P_{-t} such that the spherical projection f_t of the Legendre submanifold $P_{-t}\lambda$ is an immersion at x, and hence on a neighborhood of x. By Theorem 4.7, the corollary also holds for λ in this neighborhood of x. Since x is an arbitrary point, the corollary is proved.   □

Let $\lambda : M^{n-1} \to \Lambda^{2n-1}$ be an arbitrary Legendre submanifold. A connected submanifold S of M^{n-1} is called a *curvature surface* if at each $x \in S$, the tangent space $T_x S$ is equal to some principal space T_i. For example, if dim T_i is constant on an open subset U of M^{n-1}, then each leaf of the principal foliation T_i is a curvature surface on U. Curvature surfaces are plentiful, since the results of Reckziegel [158] and Singley [170] imply that there is an open dense subset Ω of M^{n-1} on which the multiplicities of the curvature spheres are locally constant. On Ω, each leaf of each principal foliation is a curvature surface.

It is also possible to have a curvature surface S which is not a leaf of a principal foliation, because the multiplicity of the corresponding curvature sphere is not constant on a neighborhood of S, as in the following example.

Example 4.10 (*a curvature surface which is not a leaf of a principal foliation*). Let T^2 be a torus of revolution in $\mathbf{R}^3$, and embed $\mathbf{R}^3$ into $\mathbf{R}^4 = \mathbf{R}^3 \times \mathbf{R}$. Let η be a field of unit normals to T^2 in $\mathbf{R}^3$. Let M^3 be a tube of sufficiently small radius $\varepsilon > 0$ around T^2 in $\mathbf{R}^4$, so that M^3 is a compact smooth embedded hypersurface in $\mathbf{R}^4$. The normal space to T^2 in $\mathbf{R}^4$ at a point $x \in T^2$ is spanned by $\eta(x)$ and $e_4 = (0, 0, 0, 1)$. The shape operator A_η of T^2 has two distinct principal curvatures at each point of T^2, while the shape operator A_{e_4} of T^2 is identically zero. Thus the shape operator A_ζ for the normal

$$\zeta = \cos \theta \ \eta(x) + \sin \theta \ e_4,$$

at a point $x \in T^2$, is given by

$$A_\zeta = \cos \theta \ A_{\eta(x)}.$$

From the formulas for the principal curvatures of a tube (see Cecil–Ryan [52, p. 131]), one finds that at all points of M^3 where $x_4 \neq \pm\varepsilon$, there are three distinct principal curvatures of multiplicity one, which are constant along their corresponding lines of curvature (curvature surfaces of dimension one). However, on the two tori, $T^2 \times \{\pm\varepsilon\}$, the principal curvature $\kappa = 0$ has multiplicity two. These two tori are curvature surfaces for this principal curvature, since the principal space corresponding to κ is tangent to each torus at every point. The Legendre submanifold λ induced by this embedding of M^3 in $\mathbf{R}^4$ has the same properties.

Part (e) of Corollary 4.9 has the following generalization, the proof of which is obtained by invoking the theorem of Ryan [163] mentioned in the proof of Corollary 4.9, with obvious minor modifications.

Corollary 4.11. *Suppose that S is a curvature surface of dimension $m > 1$ in a Legendre submanifold. Then the corresponding curvature sphere is constant along S.*

A hypersurface $f : M^{n-1} \to S^n$ is called a *Dupin hypersurface* if along each curvature surface, the corresponding principal curvature is constant. We generalize this to the context of Lie sphere geometry by defining a *Dupin submanifold* to be a Legendre submanifold with the property that along each curvature surface, the corresponding curvature sphere is constant. Of course, Legendre submanifolds induced by Dupin hypersurfaces in S^n are Dupin in the sense defined here. But our definition is more general, because the spherical projection of a Dupin submanifold need not be an immersion. Corollary 4.11 shows that the only curvature surfaces which must be considered in checking the Dupin property are those of dimension one. A Dupin submanifold,

$$\lambda : M^{n-1} \to \Lambda^{2n-1},$$

is said to be *proper Dupin* if the number of distinct curvature spheres is constant on M^{n-1}. The Legendre submanifold induced by the torus T^2 in Example 4.10 above is a proper Dupin submanifold. On the other hand, the Legendre submanifold induced by the tube M^3 over T^2 is Dupin, but not proper Dupin, since the number of distinct curvature spheres is not constant on M^3. By Theorem 4.7 both the Dupin and proper Dupin conditions are invariant under Lie sphere transformations. Because of this, Lie sphere geometry has proved to be a useful setting for the study of Dupinsubmanifolds.

We now begin the proof of Pinkall's theorem, Theorem 4.8. Let $\lambda : M^{n-1} \to \Lambda^{2n-1}$ be a Legendre submanifold with spherical projection f and spherical unit normal field ξ. Given $x \in M^{n-1}$, the differential df is a linear map on $T_x M^{n-1}$ that satisfies

$$df(X) \cdot f(x) = 0, \quad df(X) \cdot \xi(x) = 0,$$

for all $X \in T_x M^{n-1}$. The second equation above holds because λ is a Legendre submanifold. The differential $d\xi$ satisfies

$$d\xi(X) \cdot f(x) = 0, \quad d\xi(X) \cdot \xi(x) = 0,$$

with the first equation due to the contact condition and the fact that $f \cdot \xi = 0$. Thus, df and $d\xi$ are both linear maps from $T_x M^{n-1}$ to the vector space

$$W_x^{n-1} = (\text{Span}\{f(x), \xi(x)\})^{\perp}.$$

The first step in the proof of Theorem 4.8 is the following lemma.

Lemma 4.12. *Let $\lambda : M^{n-1} \to \Lambda^{2n-1}$ be a Legendre submanifold with spherical projection f and spherical unit normal field ξ. Let x be any point of M^{n-1}. Then for any two vectors X, Y in $T_x M^{n-1}$, we have*

$$df(X) \cdot d\xi(Y) = df(Y) \cdot d\xi(X).$$

Proof. Extend X and Y to vector fields in a neighborhood of x such that the Lie bracket $[X, Y] = 0$. First we differentiate the equation $f \cdot \xi = 0$ and obtain

$$0 = X(f \cdot \xi) = Xf \cdot \xi + f \cdot X\xi.$$

This, along with the contact condition $Xf \cdot \xi = 0$, implies that $f \cdot X\xi = 0$. We now differentiate this in the direction of Y to obtain

$$Yf \cdot X\xi + f \cdot YX\xi = 0. \tag{4.58}$$

Interchanging the roles of X and Y, we also have

$$Xf \cdot Y\xi + f \cdot XY\xi = 0. \tag{4.59}$$

Since $XY = YX$, we can subtract equation (4.59) from equation (4.58) and obtain $Xf \cdot Y\xi = Yf \cdot X\xi$, i.e., $df(X) \cdot d\xi(Y) = df(Y) \cdot d\xi(X)$.   □

Theorem 4.8 now follows from Lemma 4.13 below with

$$S = df : T_x M^{n-1} \to W_x^{n-1}, \quad T = d\xi : T_x M^{n-1} \to W_x^{n-1}.$$

The linear maps S and T satisfy requirement (a) of Lemma 4.13 because of Lemma 4.12, and they satisfy requirement (b) of Lemma 4.13 by the immersion condition in Theorem 4.2. Then by Lemma 4.13, at each $x \in M^{n-1}$, the map

$$df_t = \cos t \, df + \sin t \, d\xi$$

fails to be a bijection for at most $n - 1$ values of t in the interval $[0, \pi)$. Thus Theorem 4.8 is proved. We now prove the key lemma.

Lemma 4.13. *Let V and W be real vector spaces of dimension $n - 1$, and suppose that W has a positive definite scalar product (denoted by $\cdot$). Suppose that S and T are linear transformations from V to W that satisfy the following conditions:*

(a) $SX \cdot TY = SY \cdot TX$ *for all* $X, Y \in V$.
(b) kernel $S \cap$ kernel $T = \{0\}$.

Then there are at most $(n - 1 - \dim \text{kernel } S)$ values of $a \in \mathbf{R}$ for which the linear transformation $aS + T$ fails to be a bijection.

Proof. Let $V^* = V/\text{kernel } S$ and $W^* = \text{Image } S$. Then V^* and W^* both have the same dimension, $m = (n - 1 - \dim \text{kernel } S)$. For $X \in V$, let X^* denote the image of X under the canonical projection to V^*. For $Y \in W$, let Y^* denote the orthogonal projection of Y onto W^*. Suppose that Z is in kernel S. Then for any Y in V, condition (a) implies that

$$TZ \cdot SY = TY \cdot SZ = 0.$$

Thus TZ is orthogonal to every vector in W^*. Therefore, the mapping,

$$T^* : V^* \to W^*, \quad T^*X^* = (TX)^*,$$

is well defined. Similarly, the mapping $S^* : V^* \to W^*$ given by $S^*X^* = (SX)^*$ is well defined. Moreover, the map S^* is bijective, since its kernel is $\{0\}$, and V^* and W^* have the same dimension. We can use the bijection S^* to define a positive definite scalar product $(,)$ on V^* as follows:

$$(X^*, Y^*) = S^*X^* \cdot S^*Y^*.$$

Now define a linear transformation $L : V^* \to V^*$ by $L = S^{*-1}T^*$. Then for all X, Y in V, we have

$$SX \cdot TY = SX \cdot (TY)^* = S^*X^* \cdot T^*Y^* = S^*X^* \cdot S^*(S^{*-1}T^*Y^*)$$
$$= (X^*, S^{*-1}T^*Y^*) = (X^*, LY^*).$$

Reversing the roles of X and Y, we have

$$SY \cdot TX = (Y^*, LX^*).$$

Thus, L is self-adjoint by (a). Furthermore, for all $X \in V$, we have

$$TX \cdot TX \geq (TX)^* \cdot (TX)^* = T^*X^* \cdot T^*X^* \tag{4.60}$$
$$= S^*S^{*-1}T^*X^* \cdot S^*S^{*-1}T^*X^* = (LX^*, LX^*) = (X^*, L^2X^*).$$

Now for $X \in V$ and $a \in \mathbf{R}$, we have that X is in kernel $(aS + T)$ if and only if

$$(aSX + TX) \cdot (aSX + TX) = 0.$$

Using equation (4.60), we get

$$(aSX + TX) \cdot (aSX + TX) = a^2 SX \cdot SX + 2aSX \cdot TX + TX \cdot TX$$
$$\geq a^2(X^*, X^*) + 2a(X^*, LX^*) + (X^*, L^2X^*)$$
$$= ((aI + L)X^*, (aI + L)X^*) \geq 0,$$

where I is the identity on V^*. Hence, X is in kernel $(aS + T)$ if and only if X^* is in kernel $(aI + L)$. Since $(aI + L)$ is a symmetric transformation on a positive definite inner product space of dimension m, it fails to be a bijection for at most m values of a. For all other values of a, the vector X is in kernel $(aS + T)$ if and only if $X^* = 0$, i.e., X is in kernel S. In that case, the equation,

$$(aS + T)X = 0,$$

implies that $TX = 0$, i.e., X is in kernel T. Then condition (b) implies that $X = 0$, and thus $aS + T$ is a bijection.   □

4.5 Lie Curvatures and Isoparametric Hypersurfaces

In this section, we introduce certain natural Lie invariants of Legendre submanifolds which have been useful in the study of Dupin and isoparametric hypersurfaces.

Let $\lambda : M^{n-1} \to \Lambda^{2n-1}$ be an arbitrary Legendre submanifold. As before, we can write $\lambda = [Y_1, Y_{n+3}]$ with

$$Y_1 = (1, f, 0), \qquad Y_{n+3} = (0, \xi, 1), \tag{4.61}$$

where f and ξ are the spherical projection and spherical field of unit normals, respectively. At each point $x \in M^{n-1}$, the points on the line $\lambda(x)$ can be written in the form

$$\mu Y_1(x) + Y_{n+3}(x), \tag{4.62}$$

i.e., take μ as an inhomogeneous coordinate along the projective line $\lambda(x)$. Of course, Y_1 corresponds to $\mu = \infty$. The next two theorems give the relationship between the coordinates of the curvature spheres of λ and the principal curvatures of f, in the case where f has constant rank. In the first theorem, we assume that the spherical projection f is an immersion on M^{n-1}. By Theorem 4.8, we know that this can always be achieved locally by passing to a parallel submanifold.

Theorem 4.14. *Let* $\lambda : M^{n-1} \to \Lambda^{2n-1}$ *be a Legendre submanifold whose spherical projection* $f : M^{n-1} \to S^n$ *is an immersion. Let* Y_1 *and* Y_{n+3} *be the point sphere and great sphere maps of* λ *as in equation* (4.61). *Then the curvature spheres of* λ *at a point* $x \in M^{n-1}$ *are*

$$[K_i] = [\kappa_i Y_1 + Y_{n+3}], \quad 1 \leq i \leq g,$$

where $\kappa_1, \dots, \kappa_g$ *are the distinct principal curvatures at x of the oriented hypersurface* f *with field of unit normals* ξ. *The multiplicity of the curvature sphere* $[K_i]$ *equals the multiplicity of the principal curvature* κ_i.

Proof. Let X be a nonzero vector in $T_x M^{n-1}$. Then for any real number μ,

$$d(\mu Y_1 + Y_{n+3})(X) = (0, \mu \, df(X) + d\xi(X), 0).$$

This vector is in $\mathrm{Span}\{Y_1(x), Y_{n+3}(x)\}$ if and only if

$$\mu \, df(X) + d\xi(X) = 0,$$

i.e., μ is a principal curvature of f with corresponding principal vector X.

A second noteworthy case is when the point sphere map Y_1 is a curvature sphere of constant multiplicity m on M^{n-1}. By Corollary 4.9, the corresponding principal distribution is a foliation, and the curvature sphere map $[Y_1]$ is constant along the leaves of this foliation. Thus the map $[Y_1]$ factors through an immersion $[W_1]$ from the space of leaves V of this foliation into Q^{n+1}. We can write

$$W_1 = (1, \phi, 0),$$

where $\phi : V \to S^n$ is an immersed submanifold of codimension $m+1$. The manifold M^{n-1} is locally diffeomorphic to an open subset of the unit normal bundle B^{n-1} of the submanifold ϕ, and λ is essentially the Legendre submanifold induced by $\phi(V)$, as defined in Section 4.3. The following theorem relates the curvature spheres of λ to the principal curvatures of ϕ. Recall that the point sphere and great sphere maps for λ are given as in equation (4.37) by

$$Y_1(x, \xi) = (1, \phi(x), 0), \qquad Y_{n+3}(x, \xi) = (0, \xi, 1). \tag{4.63}$$

Theorem 4.15. *Let* $\lambda : B^{n-1} \to \Lambda^{2n-1}$ *be the Legendre submanifold induced by an immersed submanifold* $\phi(V)$ *in* S^n *of codimension* $m + 1$. *Let* Y_1 *and* Y_{n+3} *be the point sphere and great sphere maps of* λ *as in equation* (4.63). *Then the curvature spheres of* λ *at a point* $(x, \xi) \in B^{n-1}$ *are*

$$[K_i] = [\kappa_i Y_1 + Y_{n+3}], \quad 1 \leq i \leq g,$$

where $\kappa_1, \ldots, \kappa_{g-1}$ *are the distinct principal curvatures of the shape operator* A_ξ, *and* $\kappa_g = \infty$. *For* $1 \leq i \leq g - 1$, *the multiplicity of the curvature sphere* $[K_i]$ *equals the multiplicity of the principal curvature* κ_i, *while the multiplicity of* $[K_g]$ *is* m.

Proof. To find the curvature spheres of λ, we use the local trivialization of B^{n-1} given in Section 4.3, and the decomposition of the tangent space to B^{n-1} at (x, ξ) as follows:

$$T_x V \times \text{Span}\{\partial/\partial t_1, \ldots, \partial/\partial t_m\} = T_x V \times \mathbf{R}^m,$$

as in equation (4.38). First, note that $dY_1(0, Z)$ equals 0 for any $Z \in \mathbf{R}^m$, since Y_1 depends only on x. Hence, Y_1 is a curvature sphere, as expected. Furthermore, since

$$dY_1(X, 0) = (0, d\phi(X), 0)$$

is never in $\text{Span}\{Y_1(x, \xi), Y_{n+3}(x, \xi)\}$ for a nonzero $X \in T_x V$, the multiplicity of the curvature sphere Y_1 is m. If we let $[K_g] = [\kappa_g Y_1 + Y_{n+3}]$ be this curvature sphere, then we must take $\kappa_g = \infty$ to get $[Y_1]$. Using equation (4.39), we find the other curvature spheres at (x, ξ) by computing

$$d(\mu Y_1 + Y_{n+3})(X, 0) = (0, d\phi(\mu X - A_\xi X), 0).$$

From this it is clear that $[\mu Y_1 + Y_{n+3}]$ is a curvature sphere with principal vector $(X, 0)$ if and only if μ is a principal curvature of A_ξ with corresponding principal vector X.

Given these two theorems, we define a *principal curvature* of a Legendre submanifold $\lambda : M^{n-1} \to \Lambda^{2n-1}$ at a point $x \in M^{n-1}$ to be a value κ in the set $\mathbf{R} \cup \{\infty\}$ such that $[\kappa Y_1(x) + Y_{n+3}(x)]$ is a curvature sphere of λ at x, where Y_1 and Y_{n+3} are as in equation (4.61).

Remark 4.16. See Reckziegel [158] for a Riemannian treatment of the notion of principal curvatures and curvature surfaces in the case of an immersed submanifold $\phi : V \to S^n$ of codimension greater than one. In that case, Reckziegel defines a curvature surface to be a connected submanifold $S \subset V$ for which there is a parallel (with respect to the normal connection) section of the unit normal bundle $\eta : S \to B^{n-1}$ such that for each $x \in S$, the tangent space $T_x S$ is equal to some eigenspace of $A_{\eta(x)}$. The corresponding principal curvature $\kappa : S \to \mathbf{R}$ is then a smooth function on S. Pinkall [151] calls a submanifold $\phi(V)$ of codimension greater than one "Dupin" if along each curvature surface (in the sense of Reckziegel), the corresponding principal curvature is constant. One can show that Pinkall's definition is equivalent to our definition of a Dupin submanifold in the case where $\lambda : B^{n-1} \to \Lambda^{2n-1}$ is the Legendre submanifold induced by an immersed submanifold $\phi(V)$ in S^n of codimension greater than one.

The principal curvatures of a Legendre submanifold are not Lie invariant and depend on the special parametrization for λ given in equation (4.61). However, R. Miyaoka [113] pointed out that the cross-ratios of the principal curvatures are Lie invariant. In order to formulate Miyaoka's theorem, we need to introduce some notation. Suppose that β is a Lie sphere transformation. The Legendre submanifold $\beta\lambda$ has point sphere and great sphere maps given, respectively, by

$$Z_1 = (1, h, 0), \qquad Z_{n+3} = (0, \zeta, 1),$$

where h and ζ are the spherical projection and spherical field of unit normals for $\beta\lambda$. Suppose that

$$[K_i] = [\kappa_i Y_1 + Y_{n+3}], \quad 1 \leq i \leq g,$$

are the distinct curvature spheres of λ at a point $x \in M^{n-1}$. By Theorem 4.7, the points $\beta[K_i]$, $1 \leq i \leq g$, are the distinct curvature spheres of $\beta\lambda$ at x. We can write

$$\beta[K_i] = [\gamma_i Z_1 + Z_{n+3}], \quad 1 \leq i \leq g.$$

These γ_i are the principal curvatures of $\beta\lambda$ at x.

For four distinct numbers a, b, c, d in $\mathbf{R} \cup \{\infty\}$, we adopt the notation

$$[a, b; c, d] = \frac{(a-b)(d-c)}{(a-c)(d-b)} \tag{4.64}$$

for the cross-ratio of a, b, c, d. We use the usual conventions involving operations with ∞. For example, if $d = \infty$, then the expression $(d-c)/(d-b)$ evaluates to one, and the cross-ratio $[a, b; c, d]$ equals $(a-b)/(a-c)$.

Miyaoka's theorem can now be stated as follows.

Theorem 4.17. *Let* $\lambda : M^{n-1} \to \Lambda^{2n-1}$ *be a Legendre submanifold and* β *a Lie sphere transformation. Suppose that* $\kappa_1, \ldots, \kappa_g, g \geq 4$, *are the distinct principal curvatures of* λ *at a point* $x \in M^{n-1}$, *and* $\gamma_1, \ldots, \gamma_g$ *are the corresponding principal curvatures of* $\beta\lambda$ *at* x. *Then for any choice of four numbers* h, i, j, k *from the set* $\{1, \ldots, g\}$, *we have*

$$[\kappa_h, \kappa_i; \kappa_j, \kappa_k] = [\gamma_h, \gamma_i; \gamma_j, \gamma_k]. \tag{4.65}$$

Proof. The left side of equation (4.65) is the cross-ratio, in the sense of projective geometry, of the four points $[K_h], [K_i], [K_j], [K_k]$ on the projective line $\lambda(x)$. The right side of equation (4.65) is the cross-ratio of the images of these four points under β. The theorem now follows from the fact that the projective transformation β preserves the cross-ratio of four points on a line. $\square$

The cross-ratios of the principal curvatures of λ are called the *Lie curvatures* of λ. A set of related invariants for the Möbius group is obtained as follows. First, recall that a Möbius transformation is a Lie sphere transformation that takes point spheres to point spheres. Hence the transformation β in Theorem 4.17 is a Möbius transformation if and only if $\beta[Y_1] = [Z_1]$. This leads to the following corollary of Theorem 4.17.

Corollary 4.18. *Let* $\lambda : M^{n-1} \to \Lambda^{2n-1}$ *be a Legendre submanifold and* β *a Möbius transformation. Then for any three distinct principal curvatures* $\kappa_h, \kappa_i, \kappa_j$ *of* λ *at a point* $x \in M^{n-1}$, *none of which equals* ∞, *we have*

$$\Phi(\kappa_h, \kappa_i, \kappa_j) = (\kappa_h - \kappa_i)/(\kappa_h - \kappa_j) = (\gamma_h - \gamma_i)/(\gamma_h - \gamma_j), \quad (4.66)$$

where γ_h, γ_i *and* γ_j *are the corresponding principal curvatures of* $\beta\lambda$ *at the point* x.

Proof. First, note that we are using equation (4.66) to define the quantity Φ. Now since β is a Möbius transformation, the point $[Y_1]$, corresponding to $\mu = \infty$, is taken by β to the point Z_1 with coordinate $\gamma = \infty$. Since β preserves cross-ratios, we have

$$[\kappa_h, \kappa_i; \kappa_j, \infty] = [\gamma_h, \gamma_i; \gamma_j, \infty].$$

The corollary now follows since the cross-ratio on the left in the equation above equals the left side of equation (4.66), and the cross-ratio on the right above equals the right side of equation (4.66).   □

A ratio Φ of the form (4.66) is called a *Möbius curvature* of λ. Lie and Möbius curvatures have been useful in characterizing Legendre submanifolds that are Lie equivalent to Legendre submanifolds induced by isoparametric hypersurfaces in spheres.

Recall that an immersed hypersurface in a real space-form, $\mathbf{R}^n$, S^n or H^n, is said to be *isoparametric* if it has constant principal curvatures. As noted in the introduction, an isoparametric hypersurface M in $\mathbf{R}^n$ can have at most two distinct principal curvatures, and M must be an open subset of a hyperplane, hypersphere or a spherical cylinder $S^k \times \mathbf{R}^{n-k-1}$. (Levi–Civita [101] for $n = 3$ and B. Segre [169] for arbitrary n.)

In a series of four papers, Cartan [16]–[19] studied isoparametric hypersurfaces in the other space-forms In H^n, he showed that an isoparametric hypersurface can have at most two distinct principal curvatures, and it is either totally umbilic or else a standard product $S^k \times H^{n-k-1}$ in H^n (see also Ryan [164, pp. 252–253]).

In the sphere S^n, however, Cartan found examples of isoparametric hypersurfaces in S^n with $1, 2, 3$ or 4 distinct principal curvatures, and he classified isoparametric hypersurfaces with $g \leq 3$ principal curvatures. Cartan's theory was further developed by Münzner [123], who showed among other things that the number g of distinct principal curvatures of an isoparametric hypersurface must be $1, 2, 3, 4$ or 6. The major results in the theory of isoparametric hypersurfaces in spheres are described in detail in the introduction. (See also the survey article by Thorbergsson [192].)

Münzner's work shows that any connected isoparametric hypersurface in S^n can be extended to a compact, connected isoparametric hypersurface in a unique way. The following is a local Lie geometric characterization of those Legendre submanifolds that are Lie equivalent to a Legendre submanifold induced by an isoparametric hypersurface in S^n (see Cecil [33]). Recall that a line in $\mathbf{P}^{n+2}$ is called *timelike* if it contains only timelike points. This means that an orthonormal basis for the 2-plane in $\mathbf{R}_2^{n+3}$ determined by the timelike line consists of two timelike vectors. An example is the line $[e_1, e_{n+3}]$.

Theorem 4.19. *Let* $\lambda : M^{n-1} \to \Lambda^{2n-1}$ *be a Legendre submanifold with g distinct curvature spheres* $[K_1], \ldots, [K_g]$ *at each point. Then* λ *is Lie equivalent to the Legendre submanifold induced by an isoparametric hypersurface in* S^n *if and only if there exist g points* $[P_1], \ldots, [P_g]$ *on a timelike line in* $\mathbf{P}^{n+2}$ *such that*

$$\langle K_i, P_i \rangle = 0, \quad 1 \leq i \leq g.$$

Proof. If λ is the Legendre submanifold induced by an isoparametric hypersurface in S^n, then all the spheres in a family $[K_i]$ have the same radius ρ_i, where $0 < \rho_i < \pi$. By formula (2.21) of Chapter 2, p. 17, this is equivalent to the condition $\langle K_i, P_i \rangle = 0$, where

$$P_i = \sin \rho_i \, e_1 - \cos \rho_i \, e_{n+3}, \quad 1 \leq i \leq g, \tag{4.67}$$

are g points on the timelike line $[e_1, e_{n+3}]$. Since a Lie sphere transformation preserves curvature spheres, timelike lines and the polarity relationship, the same is true for any image of λ under a Lie sphere transformation.

Conversely, suppose that there exist g points $[P_1], \ldots, [P_g]$ on a timelike line ℓ such that

$$\langle K_i, P_i \rangle = 0, \quad 1 \leq i \leq g.$$

Let β be a Lie sphere transformation that maps ℓ to the line $[e_1, e_{n+3}]$. Then the curvature spheres $\beta[K_i]$ of $\beta\lambda$ are respectively orthogonal to the points $[Q_i] = \beta[P_i]$ on the line $[e_1, e_{n+3}]$. This means that the spheres corresponding to $\beta[K_i]$ have constant radius on M^{n-1}. By applying a parallel transformation, if necessary, we can arrange that none of these curvature spheres has radius zero. Then $\beta\lambda$ is the Legendre submanifold induced by an isoparametric hypersurface in S^n. $\qquad \square$

Remark 4.20. In the case where λ is Lie equivalent to the Legendre submanifold induced by an isoparametric hypersurface in S^n, one can say more about the position of the points $[P_1], \ldots, [P_g]$ on the timelike line ℓ. Münzner showed that the radii ρ_i of the curvature spheres of an isoparametric hypersurface must be of the form

$$\rho_i = \rho_1 + (i - 1)\frac{\pi}{g}, \quad 1 \leq i \leq g, \tag{4.68}$$

for some $\rho_1 \in (0, \pi/g)$. Hence, after Lie sphere transformation, the $[P_i]$ must have the form (4.67) for ρ_i as in equation (4.68).

Since the principal curvatures are constant on an isoparametric hypersurface, the Lie curvatures are also constant. By Münzner's work, the distinct principal curvatures κ_i, $1 \leq i \leq g$, of an isoparametric hypersurface must have the form

$$\kappa_i = \cot \rho_i, \tag{4.69}$$

for ρ_i as in equation (4.68). Thus the Lie curvatures of an isoparametric hypersurface can be determined. We can order the principal curvatures so that

$$\kappa_1 < \cdots < \kappa_g. \tag{4.70}$$

In the case $g = 4$, this leads to a unique Lie curvature Ψ defined by

$$\Psi = [\kappa_1, \kappa_2; \kappa_3, \kappa_4] = (\kappa_1 - \kappa_2)(\kappa_4 - \kappa_3)/(\kappa_1 - \kappa_3)(\kappa_4 - \kappa_2). \qquad (4.71)$$

The ordering of the principal curvatures implies that Ψ must satisfy $0 < \Psi < 1$. Using equations (4.69) and (4.71), one can compute that $\Psi = 1/2$ on any isoparametric hypersurface, i.e., the four curvature spheres form a *harmonic set* in the sense of projective geometry (see, for example, [166, p. 59]). There is, however, a simpler way to compute Ψ. One applies Theorem 4.15 to the Legendre submanifold induced by one of the focal submanifolds of the isoparametric hypersurface. By the work of Münzner, each isoparametric hypersurface M^{n-1} embedded in S^n has two distinct focal submanifolds, each of codimension greater than one. The hypersurface M^{n-1} is a tube of constant radius over each of these focal submanifolds. Therefore, the Legendre submanifold induced by M^{n-1} is obtained from the Legendre submanifold induced by either focal submanifold by parallel transformation. Thus, the Legendre submanifold induced by M^{n-1} has the same Lie curvature as the Legendre submanifold induced by either focal submanifold. Let $\phi : V \to S^n$ be one of the focal submanifolds. By the same calculation that yields equation (4.68), Münzner showed that if ξ is any unit normal to $\phi(V)$ at any point, then the shape operator A_ξ has three distinct principal curvatures,

$$\kappa_1 = -1, \qquad \kappa_2 = 0, \qquad \kappa_3 = 1.$$

By Theorem 4.15, the Legendre submanifold induced by ϕ has a fourth principal curvature $\kappa_4 = \infty$. Thus, the Lie curvature of this Legendre submanifold is

$$\Psi = (-1 - 0)(\infty - 1)/(-1 - 1)(\infty - 0) = 1/2.$$

We can determine the Lie curvatures of an isoparametric hypersurface M^{n-1} in S^n with $g = 6$ principal curvatures in the same way. Let $\phi(V)$ be one of the focal submanifolds of M^{n-1}. By Münzner's formula (4.68), the Legendre submanifold induced by $\phi(V)$ has six constant principal curvatures,

$$\kappa_1 = -\sqrt{3}, \quad \kappa_2 = -1/\sqrt{3}, \quad \kappa_3 = 0, \quad \kappa_4 = 1/\sqrt{3}, \quad \kappa_5 = \sqrt{3}, \quad \kappa_6 = \infty.$$

The corresponding six curvature spheres are situated symmetrically, as in Figure 4.2. There are only three geometrically distinct configurations which can obtained by choosing four of the six curvature spheres. These give the cross-ratios:

$$[\kappa_3, \kappa_4; \kappa_5, \kappa_6] = 1/3, \qquad [\kappa_2, \kappa_3; \kappa_5, \kappa_6] = 1/4, \qquad [\kappa_2, \kappa_3; \kappa_4, \kappa_6] = 1/2.$$

Of course, if a certain cross-ratio has the value Ψ, then one can obtain the values

$$\{\Psi, 1/\Psi, 1 - \Psi, 1/(1 - \Psi), (\Psi - 1)/\Psi, \Psi/(\Psi - 1)\}, \qquad (4.72)$$

by permuting the order of the spheres (see, for example, Samuel [166, p. 58]).

Returning to the case $g = 4$, one can ask what is the strength of the assumption $\Psi = 1/2$ on M^{n-1}. Since Ψ is only one function of the principal curvatures, one

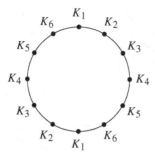

Fig. 4.2. Curvature spheres on a projective line, $g = 6$.

would not expect this assumption to classify Legendre submanifolds up to Lie equivalence. However, if one makes additional assumptions, e.g., the Dupin condition, then results can be obtained.

Miyaoka [113] proved that the assumption that Ψ is constant on a compact connected proper Dupin hypersurface M^{n-1} in S^n with four principal curvatures, together with an additional assumption regarding intersections of leaves of the various principal foliations, implies that M^{n-1} is Lie equivalent to an isoparametric hypersurface. Note Thorbergsson [190] showed that for a compact proper Dupin hypersurface in S^n with four principal curvatures, the multiplicities of the principal curvatures must satisfy $m_1 = m_2, m_3 = m_4$, when the principal curvatures are appropriately ordered (see also Stolz [177]).

Cecil, Chi and Jensen [41] used a different approach than Miyaoka to prove that if M^{n-1} is a compact connected proper Dupin hypersurface in S^n with four principal curvatures and constant Lie curvature, whose multiplicities satisfy $m_1 = m_2 \geq 1$, $m_3 = m_4 = 1$, then M^{n-1} is Lie equivalent to an isoparametric hypersurface. Thus, Miyaoka's additional assumption is not needed in that case. It remains an open question whether Miyaoka's additional assumption can be removed in the case where $m_3 = m_4$ is also allowed to be greater than one, although this has been conjectured to be true by Cecil and Jensen [45, pp. 3–4].

In the same paper [41], Cecil, Chi and Jensen also obtained a local result by showing that if a connected proper Dupin submanifold,

$$\lambda : M^{n-1} \to \Lambda^{2n-1},$$

has four distinct principal curvatures with multiplicities,

$$m_1 = m_2 \geq 1, \qquad m_3 = m_4 = 1, \tag{4.73}$$

and constant Lie curvature $\Psi = -1$, and λ is irreducible (in the sense of Pinkall [150], see Section 5.1), then λ is Lie equivalent to the Legendre submanifold induced by an isoparametric hypersurface in S^n. (Note that if the principal curvatures of an isoparametric hypersurface with four principal curvatures are ordered so that the multiplicities satisfy $m_1 = m_2, m_3 = m_4$, then the Lie curvature equals -1 instead of $1/2$, see equation (4.72)). Again the conjecture of Cecil and Jensen [45, pp. 3–4] states that this result also holds if $m_3 = m_4$ is allowed to be greater than one.

We now return to the ordering of the principal curvatures given in equation (4.70), so that the Lie curvature $\Psi = 1/2$ for an isoparametric hypersurface with four principal curvatures.

The following example of Cecil [33] shows that some additional hypotheses besides $\Psi = 1/2$ are needed to be able to conclude that a proper Dupin hypersurface with four principal curvatures is Lie equivalent to an isoparametric hypersurface. This example is a noncompact proper Dupin submanifold with $g = 4$ distinct principal curvatures and constant Lie curvature $\Psi = 1/2$, which is not Lie equivalent to an open subset of an isoparametric hypersurface with four principal curvatures in S^n. This example cannot be made compact without destroying the property that the number g of distinct curvatures spheres equals four at each point.

Let $\phi : V \to S^{n-m}$ be an embedded Dupin hypersurface in S^{n-m} with field of unit normals ξ, such that ϕ has three distinct principal curvatures,

$$\mu_1 < \mu_2 < \mu_3,$$

at each point of V. Embed S^{n-m} as a totally geodesic submanifold of S^n, and let B^{n-1} be the unit normal bundle of the submanifold $\phi(V)$ in S^n. Let $\lambda : B^{n-1} \to \Lambda^{2n-1}$ be the Legendre submanifold induced by the submanifold $\phi(V)$ in S^n. Any unit normal η to $\phi(V)$ at a point $x \in V$ can be written in the form

$$\eta = \cos\theta\,\xi(x) + \sin\theta\,\zeta,$$

where ζ is a unit normal to S^{n-m} in S^n. Since the shape operator $A_\zeta = 0$, we have

$$A_\eta = \cos\theta\,A_\xi.$$

Thus the principal curvatures of A_η are

$$\kappa_i = \cos\theta\,\mu_i, \quad 1 \le i \le 3. \tag{4.74}$$

If $\eta \cdot \xi = \cos\theta \ne 0$, then A_η has three distinct principal curvatures. However, if $\eta \cdot \xi = 0$, then $A_\eta = 0$. Let U be the open subset of B^{n-1} on which $\cos\theta > 0$, and let α denote the restriction of λ to U. By Theorem 4.15, α has four distinct curvature spheres at each point of U. Since $\phi(V)$ is Dupin in S^{n-m}, it is easy to show that α is Dupin (see the tube construction in Section 5.2 for the details). Furthermore, since $\kappa_4 = \infty$, the Lie curvature Ψ of α at a point (x, η) of U equals the Möbius curvature $\Phi(\kappa_1, \kappa_2, \kappa_3)$. Using equation (4.74), we compute

$$\Psi = \Phi(\kappa_1, \kappa_2, \kappa_3) = \frac{\kappa_1 - \kappa_2}{\kappa_1 - \kappa_3} = \frac{\mu_1 - \mu_2}{\mu_1 - \mu_3} = \Phi(\mu_1, \mu_2, \mu_3). \tag{4.75}$$

Now suppose that $\phi(V)$ is a minimal isoparametric hypersurface in S^{n-m} with three distinct principal curvatures. By Münzner's formula (4.68), these principal curvatures must have the values

$$\mu_1 = -\sqrt{3}, \qquad \mu_2 = 0, \qquad \mu_3 = \sqrt{3}.$$

On the open subset U of B^{n-1} described above, the Lie curvature of α has the constant value $1/2$ by equation (4.75). To construct an immersed proper Dupin hypersurface with four principal curvatures and constant Lie curvature $\Psi = 1/2$ in S^n, we simply take the open subset $\phi_t(U)$ of the tube of radius t around $\phi(V)$ in S^n.

It is not hard to see that this example is not Lie equivalent to an open subset of an isoparametric hypersurface in S^n with four distinct principal curvatures. Note that the point sphere map $[Y_1]$ of α is a curvature sphere of multiplicity m which lies in the linear subspace of codimension $m+1$ in $\mathbf{P}^{n+2}$ orthogonal to the space spanned by e_{n+3} and by those vectors ζ normal to S^{n-m} in S^n. This geometric fact implies that for such a vector ζ, there are only two distinct curvature spheres on each of the lines $\lambda(x, \zeta)$, since $A_\zeta = 0$ (see Theorem 4.15). On the other hand, if $\gamma : M^{n-1} \to \Lambda^{2n-1}$ is the Legendre submanifold induced by an isoparametric hypersurface in S^n with four distinct principal curvatures, then there are four distinct curvature spheres on each line $\gamma(x)$, for $x \in M^{n-1}$. Thus, no curvature sphere of γ lies in a linear subspace of codimension greater than one in $\mathbf{P}^{n+2}$, and so γ is not Lie equivalent to α. This change in the number of distinct curvature spheres at points of the form (x, ζ) is precisely why α cannot be extended to a compact proper Dupin submanifold with $g = 4$.

With regard to Theorem 4.19, α comes as close as possible to satisfying the requirements for being Lie equivalent to an isoparametric hypersurface without actually fulfilling them. The principal curvatures $\kappa_2 = 0$ and $\kappa_4 = \infty$ are constant on U. If a third principal curvature were also constant, then the constancy of Ψ would imply that all four principal curvatures were constant, and α would be the Legendre submanifold induced by an isoparametric hypersurface.

Using this same method, it is easy to construct noncompact proper Dupin hypersurfaces in S^n with $g = 4$ and $\Psi = c$, for any constant $0 < c < 1$. If $\phi(V)$ is an isoparametric hypersurface in S^{n-m} with three distinct principal curvatures, then Münzner's formula (4.68) implies that these principal curvatures must have the values

$$\mu_1 = \cot\left(\theta + \frac{2\pi}{3}\right), \qquad \mu_2 = \cot\left(\theta + \frac{\pi}{3}\right), \qquad \mu_3 = \cot\theta, \quad 0 < \theta < \frac{\pi}{3}.$$
(4.76)

Furthermore, any value of θ in $(0, \pi/3)$ can be realized by some hypersurface in a parallel family of isoparametric hypersurfaces. A direct calculation using equations (4.75) and (4.76) shows that the Lie curvature Ψ of α satisfies

$$\Psi = \Phi(\kappa_1, \kappa_2, \kappa_3) = \frac{\kappa_1 - \kappa_2}{\kappa_1 - \kappa_3} = \frac{\mu_1 - \mu_2}{\mu_1 - \mu_3} = \frac{1}{2} + \frac{\sqrt{3}}{2}\tan\left(\theta - \frac{\pi}{6}\right),$$

on U. This can assume any value c in the interval $(0, 1)$ by an appropriate choice of θ in $(0, \pi/3)$. An open subset of a tube over $\phi(V)$ in S^n is a proper Dupin hypersurface with $g = 4$ and $\Psi = \Phi = c$. Note that Φ has different values on different hypersurfaces in the parallel family of isoparametric hypersurfaces. Thus these hypersurfaces are not Möbius equivalent to each other by Corollary 4.18. This is consistent with the fact that a parallel transformation is not a Möbius transformation.

4.6 Lie Invariance of Tautness

In this section, we prove that tautness is invariant under Lie sphere transformations. A proof of this result was first given in [37]. However, the proof that we give here is due to Álvarez Paiva [2], who used functions whose level sets form a parabolic pencil of spheres rather than the usual distance functions to formulate tautness. This leads to a very natural proof of the Lie invariance of tautness.

First, we briefly review the definition and basic facts concerning taut immersions into real space forms. The reader is referred to Cecil–Ryan [52, Chapter 2] or the survey article [36] in the book [39] for more detail and additional references. For a generalization of tautness to submanifolds of arbitrary Riemannian manifolds, see Terng and Thorbergsson [189].

Let $\phi : V \to \mathbf{R}^n$ be an immersion of a compact, connected manifold V into $\mathbf{R}^n$ with dim $V < n$. For $p \in \mathbf{R}^n$, the *Euclidean distance function*, $L_p : V \to \mathbf{R}$, is defined by the formula

$$L_p(x) = |p - \phi(x)|^2.$$

If p is not a focal point of the submanifold ϕ, then L_p is a *nondegenerate function* (or a *Morse function*), i.e., all of its critical points are nondegenerate (see Milnor [110, pp. 32–38]). By the Morse inequalities, the number $\mu(L_p)$ of critical points of a nondegenerate distance function on V satisfies

$$\mu(L_p) \geq \beta(V; \mathbf{F}),$$

the sum of the **F**-Betti numbers of V for any field **F**. The immersion ϕ is said to be *taut* if there exists a field **F** such that every nondegenerate Euclidean distance function has $\beta(V; \mathbf{F})$ critical points on V. If it is necessary to distinguish the field **F**, we will say that ϕ is **F**-*taut*. The field $\mathbf{F} = \mathbf{Z}_2$ has been satisfactory for most considerations thus far, and we will use $\mathbf{Z}_2$ as the field in this section.

Tautness was first studied by Banchoff [6], who determined all taut two-dimensional surfaces in Euclidean space. Carter and West [21] introduced the term "taut immersion," and proved many basic results in the field. Among these is the fact that a taut immersion must be an embedding, since if $p \in \mathbf{R}^n$ were a double point, then the function L_p would have two absolute minima instead of just one, as required by tautness.

Tautness is invariant under Möbius transformations of $\mathbf{R}^n \cup \{\infty\}$. Further, an embedding $\phi : V \to \mathbf{R}^n$ of a compact manifold V is taut if and only if the embedding $\sigma\phi : V \to S^n$ has the property that every nondegenerate spherical distance function $d_p(q) = \cos^{-1}(p \cdot q)$ has $\beta(V; \mathbf{F})$ critical points on V, where $\sigma : \mathbf{R}^n \to S^n - \{P\}$ is stereographic projection (see Section 2.2). Since a spherical distance function d_p is essentially a Euclidean height function $\ell_p(q) = p \cdot q$, for $p, q \in S^n$, the embedding ϕ is taut if and only if the spherical embedding $\sigma\phi$ is *tight*, i.e., every nondegenerate height function ℓ_p has $\beta(V; \mathbf{F})$ critical points on V. It is often simpler to use height functions rather than spherical distance functions when studying tautness for submanifolds of S^n. (For a survey of results on tight submanifolds, see Kuiper [97]–[98], Banchoff and Kühnel [7] or Cecil–Ryan [52, Chapter 1]. For taut submanifolds of hyperbolic space, see Cecil [30], and Cecil and Ryan [48]–[49].)

In this proof of the Lie invariance of tautness, it is more convenient to consider embeddings into S^n rather than $\mathbf{R}^n$. The remarks in the paragraph above show that the two theories are essentially equivalent for embeddings of compact manifolds V.

Kuiper [96] gave a reformulation of critical point theory in terms of an injectivity condition on homology which has turned out to be very fruitful in the theory of tight and taut immersions. Let f be a nondegenerate function on a manifold V. We define the *sublevel set*

$$V_r(f) = \{x \in V \mid f(x) \le r\}, \quad r \in \mathbf{R}. \tag{4.77}$$

The next theorem, which follows immediately from Morse–Cairns [122, Theorem 2.2, p. 260] (see also Cecil–Ryan [52, Theorem 2.1 of Chapter 1]), was a key to Kuiper's formulation of these conditions.

Theorem 4.21. *Let f be a nondegenerate function on a compact, connected manifold V. For a given field $\mathbf{F}$, the number $\mu(f)$ of critical points of f equals the sum $\beta(V; \mathbf{F})$ of the $\mathbf{F}$-Betti numbers of V if and only if the map on homology,*

$$H_*(V_r(f); \mathbf{F}) \to H_*(V; \mathbf{F}), \tag{4.78}$$

induced by the inclusion $V_r(f) \subset V$ is injective for all $r \in \mathbf{R}$.

Of course, for an embedding $\phi : V \to S^n$ and a height function ℓ_p, the set $V_r(\ell_p)$, is equal to $\phi^{-1}(B)$, where B is the closed ball in S^n obtained by intersecting S^n with the half-space in $\mathbf{R}^{n+1}$ determined by the inequality $\ell_p(q) \le r$. Kuiper [99] used the continuity property of $\mathbf{Z}_2$-Čech homology to formulate tautness in terms of $\phi^{-1}(B)$, for all closed balls B in S^n, not just those centered at nonfocal points of ϕ. Thus, Kuiper proved the following theorem (see also Cecil–Ryan [52, Theorem 1.12 of Chapter 2, p. 118]).

Theorem 4.22. *Let $\phi : V \to S^n$ be an embedding of a compact, connected manifold V into S^n. Then ϕ is $\mathbf{Z}_2$-taut if and only if for every closed ball B in S^n, the induced homomorphism $H_*(\phi^{-1}(B)) \to H_*(V)$ in $\mathbf{Z}_2$-Čech homology is injective.*

The key to the approach of Álvarez Paiva [2] is to formulate tautness in terms of functions whose level sets form a parabolic pencil of unoriented spheres instead of using the usual height functions ℓ_p. Specifically, given a contact element $(p, \xi) \in T_1 S^n$, we want to define a function

$$r_{(p,\xi)} : S^n - \{p\} \to (0, \pi),$$

whose level sets are unoriented spheres in the parabolic pencil of unoriented spheres determined by (p, ξ). (We will often denote $r_{(p,\xi)}$ simply by r when the context is clear.) Every point x in $S^n - \{p\}$ lies on precisely one sphere S_x in the pencil as the spherical radius r of the spheres in the pencil varies from 0 to π. The radius $r_{(p,\xi)}(x)$ of S_x is defined implicitly by the equation

$$\cos r = x \cdot (\cos r \, p + \sin r \, \xi). \tag{4.79}$$

This equation says that x lies in the unoriented sphere S_x in the pencil with center

$$q = \cos r \, p + \sin r \, \xi, \tag{4.80}$$

and spherical radius $r \in (0, \pi)$ (see Figure 4.3). This defines a smooth function

$$r_{(p,\xi)} : S^n - \{p\} \to (0, \pi).$$

Note that the contact element $(p, -\xi)$ determines the same pencil of unoriented spheres and the function $r_{(p,-\xi)} = \pi - r_{(p,\xi)}$. Some sample values of the function $r_{(p,\xi)}$ are

$$r_{(p,\xi)}(\xi) = \pi/4, \qquad r_{(p,\xi)}(-p) = \pi/2, \qquad r_{(p,\xi)}(-\xi) = 3\pi/4.$$

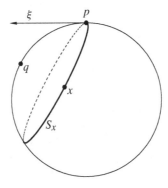

Fig. 4.3. The sphere S_x in the parabolic pencil determined by (p, ξ).

In this section, we are dealing with immersions $\phi : V \to S^n$, where V is a k-dimensional manifold with $k < n$. If $x \in V$, we say that the sphere S_x and $\phi(V)$ are *tangent* at $\phi(x)$ if

$$d\phi(T_x V) \subset T_{\phi(x)} S_x,$$

where $d\phi$ is the differential of ϕ.

The following lemma describes the critical point behavior of a function of the form $r_{(p,\xi)}$ on an immersed submanifold $\phi : V \to S^n$.

Lemma 4.23. *Let $\phi : V \to S^n$ be an immersion of a connected manifold V with $\dim V < n$ into S^n, and let $(p, \xi) \in T_1 S^n$ such that $p \notin \phi(V)$.*

(a) *A point $x_0 \in V$ is a critical point of the function $r_{(p,\xi)}$ if and only if the sphere S_{x_0} containing $\phi(x_0)$ in the parabolic pencil of unoriented spheres determined by (p, ξ) and the submanifold $\phi(V)$ are tangent at $\phi(x_0)$.*

(b) *If $r_{(p,\xi)}$ has a critical point at $x_0 \in V$, then this critical point is degenerate if and only if the sphere S_{x_0} is a curvature sphere of $\phi(V)$ at x_0.*

Proof.

(a) For any point $x_0 \in V$, there exists a sufficiently small neighborhood U of x_0 such that the restriction of ϕ to U is an embedding. We will identify U with its embedded image $\phi(U) \subset S^n$ and omit the reference to the embedding ϕ. For brevity, we will denote the function $r_{(p,\xi)}$ simply as r.

Let X be a smooth vector field tangent to V in such a neighborhood U of x_0. Then on U we compute the derivative

$$X(\cos r) = -\sin r\, X(r), \tag{4.81}$$

and $\sin r \neq 0$ for $r \in (0, \pi)$, so we see that the functions $\cos r$ and r have the same critical points on U. Using this and equation (4.79), we see that

$$X(r) = 0 \Leftrightarrow X(\cos r) = 0 \Leftrightarrow X(x \cdot (\cos r\, p + \sin r\, \xi)) = 0. \tag{4.82}$$

Then we compute

$$X(x \cdot (\cos r\, p + \sin r\, \xi)) = X \cdot (\cos r\, p + \sin r\, \xi) + x \cdot (X(\cos r)p + X(\sin r)\xi). \tag{4.83}$$

If $X(r) = 0$, then $X(\cos r) = X(\sin r) = 0$, and we have from equation (4.83) that

$$X \cdot (\cos r\, p + \sin r\, \xi) = 0. \tag{4.84}$$

Thus, if the function $r = r_{(p,\xi)}$ has a critical point at x_0, we have

$$X \cdot q = 0, \tag{4.85}$$

for all $X \in T_{x_0}V$, where $q = \cos r\, p + \sin r\, \xi$ is the center of the sphere S_{x_0} in the parabolic pencil determined by (p, ξ) containing the point x_0. The normal space to the sphere S_{x_0} in $\mathbf{R}^{n+1}$ at the point x_0 is spanned by the vectors x_0 and q. So equation (4.85), together with the fact that $X \cdot x_0 = 0$ for $X \in T_{x_0}V$, implies that the tangent space $T_{x_0}V$ is contained in the tangent space $T_{x_0}S_{x_0}$, i.e., the sphere S_{x_0} and the submanifold $\phi(V)$ are tangent at the point $\phi(x_0)$.

Conversely, if the tangent space $T_{x_0}V$ is contained in the tangent space $T_{x_0}S_{x_0}$, then $X(r) = 0$ for all $X \in T_{x_0}V$, because $S_{x_0} - \{p\}$ is a level set of the function r in S^n. Thus, r has a critical point at x_0.

(b) We now want to compute the Hessian of the function r at a critical point $x_0 \in V$. Let X and Y be smooth vector fields tangent to V on the neighborhood U of x_0 in V that was used in part (a). Then the value $H(X, Y)$ of the Hessian of r at x_0 equals $YX(r)$ at x_0. We first note that

$$YX(\cos r) = Y(-\sin r \ X(r)) = Y(-\sin r)X(r) - \sin r \ YX(r). \qquad (4.86)$$

At the critical point x_0, we have $X(r) = 0$, and thus

$$YX(\cos r) = -\sin r \ YX(r). \qquad (4.87)$$

On the other hand, by differentiating equation (4.79), we get

$$X(\cos r) = X \cdot (\cos r \ p + \sin r \ \xi) + x \cdot (X(\cos r)p + X(\sin r)\xi). \qquad (4.88)$$

Then by differentiating equation (4.88), we have

$$YX(\cos r) = D_Y X \cdot (\cos r \ p + \sin r \ \xi) + X \cdot (Y(\cos r)p + Y(\sin r)\xi) \qquad (4.89)$$
$$+ Y \cdot (X(\cos r)p + X(\sin r)\xi) + x \cdot (YX(\cos r)p + YX(\sin r)\xi),$$

where D is the Euclidean covariant derivative on $\mathbf{R}^{n+1}$. At the critical point x_0, we have

$$X(\cos r) = X(\sin r) = Y(\cos r) = Y(\sin r) = 0, \qquad (4.90)$$

and so equation (4.89) becomes

$$YX(\cos r) = D_Y X \cdot (\cos r \ p + \sin r \ \xi) + x_0 \cdot (YX(\cos r)p + YX(\sin r)\xi). \quad (4.91)$$

We now examine the two terms on the right side of equation (4.91) separately. Note that at the critical point x_0, the vector $q = \cos r \ p + \sin r \ \xi$ lies in the normal space to $\phi(V)$ at x_0, and so

$$q = \cos r \ x_0 + \sin r \ N, \qquad (4.92)$$

where N is a unit normal vector to $\phi(V)$ at x_0. Thus we can write the covariant derivative $D_Y X$ as

$$D_Y X = \nabla_Y X + \alpha(X, Y) - (X \cdot Y)x_0, \qquad (4.93)$$

where ∇ is the Levi–Civita connection on $\phi(V)$ induced from D, α is the second fundamental form of $\phi(V)$ in S^n, and the term $-(X \cdot Y)x_0$ is the second fundamental form of the sphere S^n as a hypersurface in $\mathbf{R}^{n+1}$. Then using equation (4.92) we obtain

$$D_Y X \cdot q = (\nabla_Y X + \alpha(X, Y) - (X \cdot Y)x_0) \cdot q \qquad (4.94)$$
$$= \alpha(X, Y) \cdot (\cos r \ x_0 + \sin r \ N) - (X \cdot Y)x_0 \cdot (\cos r \ x_0 + \sin r \ N)$$
$$= \sin r \ A_N X \cdot Y - \cos r \ X \cdot Y,$$

since $\nabla_Y X$ is orthogonal to q, $\alpha(X, Y) \cdot x_0 = 0$, $N \cdot x_0 = 0$, and

$$\alpha(X, Y) \cdot N = A_N X \cdot Y,$$

where A_N is the shape operator determined by the unit normal N to $\phi(V)$ at x_0.

Next we consider the second term on the right side of equation (4.91),

$$x_0 \cdot (YX(\cos r)p + YX(\sin r)\xi). \tag{4.95}$$

Note that at the critical point x_0 we have $X(r) = Y(r) = 0$, and so

$$YX(\cos r) = Y(-\sin r\ X(r)) = Y(-\sin r)X(r) - \sin r\ YX(r) \tag{4.96}$$
$$= -\sin r\ YX(r),$$

and similarly

$$YX(\sin r) = \cos r\ YX(r). \tag{4.97}$$

Thus we have

$$YX(\cos r)p + YX(\sin r)\xi = -\sin r\ YX(r)p + \cos r\ YX(r)\xi \tag{4.98}$$
$$= YX(r)(-\sin r\ p + \cos r\ \xi).$$

So the term in equation (4.95) above is

$$x_0 \cdot (YX(\cos r)p + YX(\sin r)\xi) = YX(r)(x_0 \cdot (-\sin r\ p + \cos r\ \xi)). \tag{4.99}$$

From equations (4.91), (4.94) and (4.99), we have

$$YX(\cos r) = \sin r A_N X \cdot Y - \cos r X \cdot Y + YX(r)(x_0 \cdot (-\sin r\ p + \cos r\ \xi)). \tag{4.100}$$

Using equations (4.87) and (4.100), we get

$$(-\sin r - (x_0 \cdot (-\sin r\ p + \cos r\ \xi)))YX(r) = \sin r A_N X \cdot Y - \cos r X \cdot Y. \tag{4.101}$$

Denote the coefficient of $YX(r)$ in equation (4.101) by

$$C = -\sin r - (x_0 \cdot (-\sin r\ p + \cos r\ \xi)). \tag{4.102}$$

Note that the vector

$$v = -\sin r\ p + \cos r\ \xi \tag{4.103}$$

is tangent at the point q to the geodesic in S^n from p with initial tangent vector ξ (see Figure 4.4).

On the sphere S_{x_0} centered at q, the function $x \cdot v$ takes its minimum value at p where it is equal to $-\sin r$. Thus

$$x_0 \cdot (-\sin r\ p + \cos r\ \xi) > -\sin r, \tag{4.104}$$

since $x_0 \in S_{x_0}$ and $x_0 \neq p$. So the term

$$\sin r + x_0 \cdot (-\sin r\ p + \cos r\ \xi) > 0, \tag{4.105}$$

and the coefficient C in equation (4.102) is negative. Thus we have from equation (4.101),

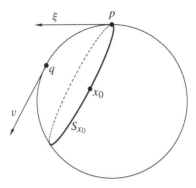

Fig. 4.4. The vector v tangent to the geodesic at q.

$$YX(r) = (1/C)(\sin r\ (A_N X \cdot Y) - \cos r\ (X \cdot Y)) \qquad (4.106)$$
$$= (1/C)(\sin r\ A_N - \cos r\ I)X \cdot Y.$$

Thus the Hessian $H(X, Y)$ of r at x_0 is degenerate if and only if there is a nonzero vector $X \in T_{x_0} V$ such that

$$(\sin r\ A_N - \cos r\ I)X = 0, \qquad (4.107)$$

that is,

$$A_N X = \cot r\ X. \qquad (4.108)$$

This equation holds if and only if $\cot r$ is an eigenvalue of A_N with corresponding principal vector X, i.e., the point $q = \cos r\ x_0 + \sin r\ N$ is a focal point of $\phi(V)$ at x_0, and the corresponding sphere S_{x_0} is a curvature sphere of $\phi(V)$ at x_0. □

Next we show that except for (p, ξ) in a set of measure zero in $T_1 S^n$, the function $r_{(p,\xi)}$ is a Morse function on $\phi(V)$. This is accomplished by using Sard's theorem in a manner similar to the usual proof that a generic distance function is a Morse function on $\phi(V)$ (see, for example, Milnor [110, pp. 32–38]). More specifically, from Lemma 4.23 we know that the function $r_{(p,\xi)}$, for $p \notin \phi(V)$, is a Morse function on $\phi(V)$ unless the parabolic pencil of unoriented spheres determined by (p, ξ) contains a curvature sphere of $\phi(V)$. We now show that the set of (p, ξ) in $T_1 S^n$ such that the parabolic pencil determined by (p, ξ) contains a curvature sphere of $\phi(V)$ has measure zero in $T_1 S^n$.

Let B^{n-1} denote the unit normal bundle of the submanifold $\phi(V)$ in S^n. Note that in the case where $\phi(V)$ is a hypersurface, B^{n-1} is a two-sheeted covering of V. We first recall the *normal exponential map*

$$q : B^{n-1} \times (0, \pi) \to S^n,$$

defined as follows. For a point (x, N) in B^{n-1} and $r \in (0, \pi)$, we define

$$q((x, N), r) = \cos r \, x + \sin r \, N. \tag{4.109}$$

Next we define a $(2n - 1)$-dimensional manifold W^{2n-1} by

$$W^{2n-1} = \{((x, N), r, \eta) \in B^{n-1} \times (0, \pi) \times S^n \mid \eta \cdot q((x, N), r) = 0\}. \tag{4.110}$$

The manifold W^{2n-1} is a fiber bundle over $B^{n-1} \times (0, \pi)$ with fiber diffeomorphic to S^{n-1}. For each point $((x, N), r) \in B^{n-1} \times (0, \pi)$, the fiber consists of all unit vectors η in $\mathbf{R}^{n+1}$ that are tangent to S^n at the point $q((x, N), r)$.
    We define a map

$$F : W^{2n-1} \to T_1 S^n \tag{4.111}$$

by

$$F((x, N), r, \eta) = (\cos r \, q + \sin r \, \eta, \ \sin r \, q - \cos r \, \eta), \tag{4.112}$$

where $q = q((x, N), r)$ is defined in equation (4.109). We will now show that if the parabolic pencil of unoriented spheres determined by $(p, \xi) \in T_1 S^n$ contains a curvature sphere of $\phi(V)$, then (p, ξ) is a critical value of F. Since the set of critical values of F has measure zero by Sard's theorem (see, for example, Milnor [110, p. 33]), this will give the desired conclusion.

Lemma 4.24. *Let* $\phi : V \to S^n$ *be an immersion of a connected manifold* V *with* $\dim V < n$ *into* S^n, *and let* B^{n-1} *be the unit normal bundle of* $\phi(V)$. *Define*

$$F : W^{2n-1} \to T_1 S^n,$$

as in equation (4.112). If the parabolic pencil of unoriented spheres determined by (p, ξ) *in* $T_1 S^n$ *contains a curvature sphere of* $\phi(V)$, *then* (p, ξ) *is a critical value of* F. *Thus, the set of such* (p, ξ) *has measure zero in* $T_1 S^n$.

Proof. Suppose that $(p, \xi) \in T_1 S^n$ is such that the sphere S with center

$$q_0 = \cos r_0 \, p + \sin r_0 \, \xi, \tag{4.113}$$

and radius $r_0 \in (0, \pi)$ is a curvature sphere of $\phi(V)$ at $x_0 \in V$. Then there is a unit normal N_0 to $\phi(V)$ at $\phi(x_0)$ such that

$$q_0 = \cos r_0 \, x_0 + \sin r_0 \, N_0, \tag{4.114}$$

and $\cot r_0$ is an eigenvalue of the shape operator A_{N_0} with corresponding nonzero principal vector X such that

$$A_{N_0} X = \cot r_0 \, X. \tag{4.115}$$

(Here, as before, we suppress the notation for the immersion ϕ and write x_0 instead of $\phi(x_0)$ in equation (4.114).) Note that if we take

$$\eta_0 = \sin r_0 \, p - \cos r_0 \, \xi,$$

then using equation (4.113), we get $\eta_0 \cdot q_0 = 0$, and

$$p = \cos r_0 \, q_0 + \sin r_0 \, \eta_0, \quad \xi = \sin r_0 \, q_0 - \cos r_0 \, \eta_0,$$

so that from equation (4.114), we have

$$(p, \xi) = F((x_0, N_0), r_0, \eta_0).$$

We now want to show that $((x_0, N_0), r_0, \eta_0)$ is a critical point of F, and thus (p, ξ) is a critical value of F. To compute the differential dF at $((x_0, N_0), r_0, \eta_0)$, we need to put local coordinates on a neighborhood of this point in W^{2n-1}. First, we choose local coordinates on the unit normal bundle B^{n-1} in a neighborhood of the point (x_0, N_0) in B^{n-1} in the following way. Suppose that V has dimension $k \leq n - 1$. Let U be a local coordinate neighborhood of x_0 in V with coordinates $(u_1, \ldots, u_k)$ such that x_0 has coordinates $(0, \ldots, 0)$. Choose orthonormal normal vector fields,

$$N_0, N_1, \ldots, N_{n-1-k},$$

on U such that the vector field N_0 agrees with the given vector N_0 at x_0, and $\nabla_X^{\perp} N_0 = 0$ at x_0, where $\nabla^{\perp}$ is the connection in the normal bundle to $\phi(V)$, for X as in equation (4.115). If $x \in U$ and N is a unit normal vector to $\phi(V)$ at $\phi(x)$ with $N \cdot N_0 > 0$, then we can write

$$N = \left(1 - \sum_{i=1}^{n-1-k} s_i^2\right)^{1/2} N_0 + s_1 N_1 + \cdots + s_{n-1-k} N_{n-1-k}, \quad \sum_{i=1}^{n-1-k} s_i^2 < 1.$$
$$(4.116)$$

Thus $(u_1, \ldots, u_k, s_1, \ldots, s_{n-1-k})$ are local coordinates on an open set O in the unit normal bundle B^{n-1} over the open set $U \subset V$, and (x_0, N_0) has coordinates $(0, \ldots, 0)$. (In the case where V has codimension one in S^n, just use the coordinates $(u_1, \ldots, u_{n-1})$ from U, since B^{n-1} is a 2-sheeted covering of U.) Therefore, any tangent vector to B^{n-1} at a point (x, N) can be written in the form (Y, W), where $Y \in T_x V$ and W is a linear combination of $\{\partial/\partial s_1, \ldots, \partial/\partial s_{n-1-k}\}$.

Next we wish to get local coordinates on the S^{n-1}-fiber near the vector η_0 at q_0. Let $\{E_1, \ldots, E_n\}$ be a local orthonormal frame of tangent vectors to S^n in a neighborhood of the point q_0 defined in equation (4.113) such that $E_n(q_0) = \eta_0$. Then we define for $((x, N), r)$ near $((x_0, N_0), r_0)$ in $B^{n-1} \times (0, \pi)$,

$$\eta((x, N), r, (t_1, \ldots, t_{n-1})) = t_1 E_1(q) + \cdots + \left(1 - \sum_{i=1}^{n-1} t_i^2\right)^{1/2} E_n(q), \quad (4.117)$$

where $q = q((x, N), r)$ is defined in equation (4.109), and $\sum_{i=1}^{n-1} t_i^2 < 1$. Thus we have local coordinates,

$$(u_1, \ldots, u_k, s_1, \ldots, s_{n-1-k}, r, t_1, \ldots, t_{n-1}),$$

on a neighborhood of the point $((x_0, N_0), r_0, \eta_0)$ in W^{2n-1}, and the point $((x_0, N_0),$ $r_0, \eta_0)$ has coordinates $(0, \ldots, 0, r_0, 0, \ldots, 0)$.

We now want to calculate the differential dF of the vector $((X, 0), 0, 0)$ tangent to W^{2n-1} at the point $((x_0, N_0), r_0, \eta_0)$. We begin by computing the differential $dq((X, 0), 0)$ at the point $((x_0, N_0), r_0)$, where q is given by equation (4.109). Let $\gamma(t)$ be a curve in U with $\gamma(0) = x_0$ and initial tangent vector $\gamma'(0) = X$. Then $(\gamma(t), N_0(\gamma(t)))$ is a curve in B^{n-1} with initial tangent vector $(X, 0)$. So at $((x_0, N_0), r_0)$, we have that $dq((X, 0), 0)$ is the initial tangent vector to the curve

$$\cos r_0 \ \gamma(t) + \sin r_0 \ N_0(\gamma(t)), \tag{4.118}$$

and so

$$dq((X, 0), 0) = \cos r_0 \ X + \sin r_0 \ D_X N_0. \tag{4.119}$$

Since $X \cdot N_0 = 0$, we have $D_X N_0 = \tilde{\nabla}_X N_0$, where $\tilde{\nabla}$ is the Levi–Civita connection on S^n, and we know that

$$\tilde{\nabla}_X N_0 = -A_{N_0} X + \nabla_X^{\perp} N_0. \tag{4.120}$$

We have chosen N_0 so that $\nabla_X^{\perp} N_0 = 0$, so by equation (4.115), we have

$$\tilde{\nabla}_X N_0 = -A_{N_0} X = -\cot r_0 \ X. \tag{4.121}$$

Thus by equations (4.119) and (4.121), we have

$$dq((X, 0), 0) = \cos r_0 \ X + \sin r_0 (-\cot r_0 \ X) = 0. \tag{4.122}$$

Next we want to compute the differential $d\eta((X, 0), 0, 0)$ at $((x_0, N_0), r_0, \eta_0)$, for η as defined in equation (4.117). Note that the coordinates $(t_1, \ldots, t_{n-1})$ are $(0, \ldots, 0)$ at $((x_0, N_0), r_0, \eta_0)$, since $E_n(q_0) = \eta_0$.

From equations (4.117) and (4.122), we see that

$$d\eta((X, 0), 0, 0) = dE_n(dq((X, 0), 0)) = 0, \tag{4.123}$$

at the point $((x_0, N_0), r_0, \eta_0)$, since $dq((X, 0), 0) = 0$. Thus from equations (4.122) and (4.123), we have

$$dF((X, 0), 0, 0) = (\cos r_0 \ dq((X, 0), 0) + \sin r_0 \ d\eta((X, 0), 0, 0), \tag{4.124}$$
$$\sin r_0 \ dq((X, 0), 0) - \cos r_0 \ d\eta((X, 0), 0, 0)) = (0, 0).$$

Thus $((x_0, N_0), r_0, \eta_0)$ is a critical point of F, and the contact element $(p, \xi) = F((x_0, N_0), r_0, \eta_0)$ is a critical value of F. This shows that every contact element whose corresponding parabolic pencil contains a curvature sphere of $\phi(V)$ is a critical value of F. By Sard's theorem, the set of critical values of the map F has measure zero in $T_1 S^n$, and so the set of contact elements (p, ξ) in $T_1 S^n$ whose parabolic pencil contains a curvature sphere of $\phi(V)$ has measure zero. □

As a consequence of Lemmas 4.23 and 4.24, we have the following corollary.

Corollary 4.25. *Let $\phi : V \to S^n$ be an immersion of a connected manifold V with* $\dim V < n$ *into* S^n. *For almost all* $(p, \xi) \in T_1 S^n$, *the function* $r_{(p,\xi)}$ *is a Morse function on* V.

Proof. By Lemma 4.23, the function $r_{(p,\xi)}$ is a Morse function on V if and only if $p \notin \phi(V)$ and the parabolic pencil of unoriented spheres determined by (p, ξ) does not contain a curvature sphere of $\phi(V)$. The set of (p, ξ) such that $p \in \phi(V)$ has measure zero, since $\phi(V)$ is a submanifold of codimension at least one in S^n. The set of (p, ξ) such that the parabolic pencil determined by (p, ξ) contains a curvature sphere of $\phi(V)$ has measure zero by Lemma 4.24. Thus, except for (p, ξ) in the set of measure zero obtained by taking the union of these two sets, the function $r_{(p,\xi)}$ is a Morse function on V. $\qquad\square$

We are now ready to give a definition of tautness for Legendre submanifolds in Lie sphere geometry. Recall the diffeomorphism from $T_1 S^n$ to the space Λ^{2n-1} of lines on the Lie quadric Q^{n+1} given by equations (4.8) and (4.9),

$$(p, \xi) \mapsto [(1, p, 0), (0, \xi, 1)] = \ell \in \Lambda^{2n-1}. \qquad (4.125)$$

Under this correspondence, an oriented sphere S in S^n belongs to the parabolic pencil of oriented spheres determined by $(p, \xi) \in T_1 S^n$ if and only if the point $[k]$ in Q^{n+1} corresponding to S lies on the line ℓ. Thus, the parabolic pencil of oriented spheres determined by a contact element (p, ξ) contains a curvature sphere S of a Legendre submanifold $\lambda : B^{n-1} \to \Lambda^{2n-1}$ if and only if the corresponding line ℓ contains the point $[k]$ corresponding to S. We now define the notion of tautness for compact Legendre submanifolds as follows. Here by "almost every," we mean except for a set of measure zero.

Definition 4.26. A compact, connected Legendre submanifold

$$\lambda : B^{n-1} \to \Lambda^{2n-1}$$

is said to be *Lie-taut* if for almost every line ℓ on the Lie quadric Q^{n+1}, the number of points $x \in B^{n-1}$ such that $\lambda(x)$ intersects ℓ is $\beta(B^{n-1}; \mathbf{Z}_2)/2$, i.e., one-half the sum of the $\mathbf{Z}_2$-Betti numbers of B^{n-1}.

Equivalently, this definition says that for almost every contact element (p, ξ) in $T_1 S^n$, the number of points $x \in B^{n-1}$ such that the contact element corresponding to $\lambda(x)$ is in oriented contact with some sphere in the parabolic pencil of oriented spheres determined by (p, ξ) is $\beta(B^{n-1}; \mathbf{Z}_2)/2$.

The property of Lie-tautness is clearly invariant under Lie sphere transformations, i.e., if $\lambda : B^{n-1} \to \Lambda^{2n-1}$ is Lie-taut and α is a Lie sphere transformation, then the Legendre submanifold $\alpha\lambda : B^{n-1} \to \Lambda^{2n-1}$ is also Lie-taut. This follows from the fact that the line $\lambda(x)$ intersects a line ℓ if and only if the line $\alpha(\lambda(x))$ intersects the line $\alpha(\ell)$, and α maps the complement of a set of measure zero in Λ^{2n-1} to the complement of a set of measure zero in Λ^{2n-1}.

Remark 4.27. The factor of one-half in the definition comes from the fact that Lie sphere geometry deals with oriented contact and not just unoriented tangency, as is made clear in the proof of the following theorem. Note here that if $\phi : V \to S^n$ is an embedding of a compact, connected manifold V into S^n and B^{n-1} is the unit normal bundle of $\phi(V)$, then the *Legendre lift* of ϕ is defined to be the Legendre submanifold $\lambda : B^{n-1} \to \Lambda^{2n-1}$ given by

$$\lambda(x, N) = [(1, \phi(x), 0), (0, N, 1)], \tag{4.126}$$

where N is a unit normal vector to $\phi(V)$ at $\phi(x)$. If V has dimension $n - 1$, then B^{n-1} is a two-sheeted covering of V. If V has dimension less than $n - 1$, then B^{n-1} is diffeomorphic to a tube W^{n-1} of sufficiently small radius over $\phi(V)$ so that W^{n-1} is an embedded hypersurface in S^n. In either case,

$$\beta(B^{n-1}; \mathbf{Z}_2) = 2\beta(V; \mathbf{Z}_2).$$

This is obvious in the case where V has dimension $n-1$, and it was proved by Pinkall [151] in the case where V has dimension less than $n - 1$.

Since Lie-tautness is invariant under Lie sphere transformations, the following theorem establishes that tautness is Lie invariant. Recall that a taut immersion $\phi : V \to S^n$ must in fact be an embedding.

Theorem 4.28. *Let $\phi : V \to S^n$ be an embedding of a compact, connected manifold V with* $\dim V < n$ *into* S^n. *Then $\phi(V)$ is a taut submanifold in S^n if and only if the Legendre lift $\lambda : B^{n-1} \to \Lambda^{2n-1}$ of ϕ is Lie-taut.*

Proof. Suppose that $\phi(V)$ is a taut submanifold in S^n, and let

$$\lambda : B^{n-1} \to \Lambda^{2n-1}$$

be the Legendre lift of ϕ. Let $(p, \xi) \in T_1 S^n$ such that $p \notin \phi(V)$ and such that the parabolic pencil of unoriented spheres determined by (p, ξ) does not contain a curvature sphere of $\phi(V)$. By Lemma 4.24, the set of such (p, ξ) is the complement of a set of measure zero in $T_1 S^n$. For such (p, ξ), the function $r_{(p,\xi)}$ is a Morse function on V, and the sublevel set

$$V_s(r_{(p,\xi)}) = \{x \in V \mid r_{(p,\xi)}(x) \le s\} = \phi(V) \cap B, \quad 0 < s < \pi, \tag{4.127}$$

is the intersection of $\phi(V)$ with a closed ball $B \subset S^n$. By tautness and Theorem 4.22, the map on $\mathbf{Z}_2$-Čech homology,

$$H_*(V_s(r_{(p,\xi)})) = H_*(\phi^{-1}(B)) \to H_*(V), \tag{4.128}$$

is injective for every $s \in \mathbf{R}$, and so by Theorem 4.21, the function $r_{(p,\xi)}$ has $\beta(V; \mathbf{Z}_2)$ critical points on V.

By Lemma 4.23, a point $x \in V$ is a critical point of $r_{(p,\xi)}$ if and only if the unoriented sphere S_x in the parabolic pencil determined by (p, ξ) containing x is tangent to $\phi(V)$ at $\phi(x)$. At each such point x, exactly one contact element $(x, N) \in B^{n-1}$ is in oriented contact with the oriented sphere $\tilde{S}_x$ through x in the parabolic pencil of oriented spheres determined by (p, ξ). Thus, the number of critical points of $r_{(p,\xi)}$ on V equals the number of points $(x, N) \in B^{n-1}$ such that (x, N) is in oriented contact with an oriented sphere in the parabolic pencil of oriented spheres determined by (p, ξ).

Thus there are

$$\beta(V; \mathbf{Z}_2) = \beta(B^{n-1}; \mathbf{Z}_2)/2$$

points $(x, N) \in B^{n-1}$ such that (x, N) is in oriented contact with an oriented sphere in the parabolic pencil of oriented spheres determined by (p, ξ). This means that there are $\beta(B^{n-1}; \mathbf{Z}_2)/2$ points $(x, N) \in B^{n-1}$ such that the line $\lambda(x, N)$ intersects the line ℓ on Q^{n+1} corresponding to the contact element (p, ξ). Since this true for almost every $(p, \xi) \in T_1 S^n$, the Legendre lift λ of ϕ is Lie-taut.

Conversely, suppose that the Legendre lift $\lambda : B^{n-1} \to \Lambda^{2n-1}$ of ϕ is Lie-taut. Then for all $(p, \xi) \in T_1 S^n$ except for a set Z of measure zero, the number of points $(x, N) \in B^{n-1}$ that are in oriented contact with some sphere in the parabolic pencil of oriented spheres determined by (p, ξ) is $\beta(B^{n-1}; \mathbf{Z}_2)/2 = \beta(V; \mathbf{Z}_2)$. This means that the corresponding function $r_{(p,\xi)}$ has $\beta(V; \mathbf{Z}_2)$ critical points on V. By Theorem 4.21, this implies that for a closed ball $B \subset S^n$ such that $\phi^{-1}(B) = V_s(r_{(p,\xi)})$ for $(p, \xi) \notin Z$ and $s \in \mathbf{R}$, the map on homology,

$$H_*(\phi^{-1}(B)) \to H_*(V), \tag{4.129}$$

is injective. On the other hand, if B is a closed ball corresponding to a sublevel set of $r_{(p,\xi)}$ for $(p, \xi) \in Z$, then since Z has measure zero, one can produce a nested sequence,

$$\{B_i\}, \quad i = 1, 2, 3, \ldots,$$

of closed balls (coming from $r_{(p,\xi)}$ for $(p, \xi) \notin Z$) satisfying

$$\phi^{-1}(B_i) \supset \phi^{-1}(B_{i+1}) \supset \cdots \supset \cap_{j=1}^{\infty} \phi^{-1}(B_j) = \phi^{-1}(B), \tag{4.130}$$

for $i = 1, 2, 3, \ldots$, such that the homomorphism in $\mathbf{Z}_2$-homology,

$$H_*(\phi^{-1}(B_i)) \to H_*(V) \text{ is injective for } i = 1, 2, 3, \ldots. \tag{4.131}$$

If equations (4.130) and (4.131) are satisfied, then the map

$$H_*(\phi^{-1}(B_i)) \to H_*(\phi^{-1}(B_j)) \text{ is injective for all } i > j. \tag{4.132}$$

The continuity property of Čech homology (see Eilenberg–Steenrod [65, p. 261]) says that

$$H_*(\phi^{-1}(B)) = \varprojlim_{i \to \infty} H_*(\phi^{-1}(B_i)).$$

Equation (4.132) and Eilenberg–Steenrod's [65, Theorem 3.4, p. 216] on inverse limits imply that the map

$$H_*(\phi^{-1}(B)) \to H_*(\phi^{-1}(B_i))$$

is injective for each i. Thus, from equation (4.131), we get that the map

$$H_*(\phi^{-1}(B)) \to H_*(V)$$

is also injective. Since this holds for all closed balls B in S^n, the embedding $\phi(V)$ is taut by Theorem 4.22. (See Kuiper [98] or Cecil–Ryan [52, Theorem 5.4, pp. 25–26] for an example of this type of Čech homology argument.) $\square$

Another formulation of the Lie invariance of tautness is the following corollary.

Corollary 4.29. *Let $\phi : V \to S^n$ and $\psi : V \to S^n$ be two embeddings of a compact, connected manifold V with $\dim V < n$ into S^n, such that their corresponding Legendre lifts are Lie equivalent. Then ϕ is taut if and only if ψ is taut.*

Proof. Since the Legendre lifts of ϕ and ψ are Lie equivalent, the unit normal bundles of $\phi(V)$ and $\psi(V)$ must be diffeomorphic, and we will denote them both by B^{n-1}. Now let $\lambda : B^{n-1} \to \Lambda^{2n-1}$ and $\mu : B^{n-1} \to \Lambda^{2n-1}$ be the Legendre lifts of ϕ and ψ, respectively. By Theorem 4.28, ϕ is taut if and only if λ is Lie-taut, and ψ is taut if and only if μ is Lie-taut. Further, since λ and μ are Lie equivalent, λ is Lie-taut if and only if μ is Lie-taut, so it follows that ϕ is taut if and only if ψ is taut. $\square$

As we noted in the introduction, every taut submanifold of a real space-form is Dupin, although not necessarily proper Dupin. (Pinkall [151], and Miyaoka [112], independently, for hypersurfaces. See also Cecil–Ryan [52, p. 195].) Thorbergsson [190] showed that a compact proper Dupin hypersurface embedded in $\mathbf{R}^n$ is taut. Pinkall [151] then extended this result to compact submanifolds of higher codimension for which the number of distinct principal curvatures is constant on the unit normal bundle. An important open question is whether Dupin implies taut without the assumption that the number of distinct principal curvatures is constant on the unit normal bundle.

It is well known that the compact focal submanifolds of a taut hypersurface in $\mathbf{R}^n$ need not be taut. For example, one focal submanifold of a nonround cyclide of Dupin in $\mathbf{R}^3$ is an ellipse. This is tight but not taut. More generally, Buyske [13] used Lie sphere geometric techniques to show that if a hypersurface M^{n-1} in $\mathbf{R}^n$ is Lie equivalent to an isoparametric hypersurface in S^n, then each focal submanifold of M^{n-1} is tight in $\mathbf{R}^n$.

4.7 Isoparametric Hypersurfaces of FKM-type

In this section, we describe a construction due to Ferus, Karcher, and Münzner [73] which gives an infinite collection of families of isoparametric hypersurfaces with four principal curvatures. In fact, this construction gives all known examples of isoparametric hypersurfaces with four principal curvatures with the exception of two homogeneous families. The hypersurfaces produced by this construction are now known as isoparametric hypersurfaces of *FKM-type*.

We will follow the paper of Ferus, Karcher, and Münzner closely, although we will not prove everything that they prove concerning these examples, and the reader is referred to their paper [73] for a detailed study of many aspects of these hypersurfaces.

Pinkall and Thorbergsson [152] later gave an alternate geometric construction of the FKM-hypersurfaces (see Section 4.8) which they then modified to produce examples of compact proper Dupin hypersurfaces with four principal curvatures and nonconstant Lie curvature. Q.-M. Wang [197]–[198] provided more information on isoparametric hypersurfaces and the topology of the FKM-hypersurfaces. In a series of papers, Dorfmeister and Neher [58]–[62] gave an algebraic treatment of the theory of isoparametric hypersurfaces, in general, and the FKM-hypersurfaces, in particular, from the point of view of triple systems.

Before we give the construction of Ferus, Karcher, and Münzner, we first need to recall some fundamental results on isoparametric hypersurfaces due to Münzner [123] (see also [52, Chapter 3]).

As in Section 4.5, an oriented hypersurface $f : M^{n-1} \to S^n \subset \mathbf{R}^{n+1}$ with field of unit normals $\xi : M^{n-1} \to S^n$ is said to be *isoparametric* if it has constant principal curvatures. As in Section 4.4, for each real number t, we define a map $f_t : M^{n-1} \to S^n$ by

$$f_t = \cos t \, f + \sin t \, \xi. \tag{4.133}$$

A point $p = f_t(x)$ is called a *focal point of multiplicity* $m > 0$ of f at x if the nullity of df_t is equal to m at x. This happens if and only if $\cot t$ is a principal curvature of multiplicity m at x.

If $f : M^{n-1} \to S^n$ is an isoparametric hypersurface with g distinct principal curvatures $\cot \theta_1, \ldots, \cot \theta_g$, and $\cot t$ is not a principal curvature of f, then f_t is also an isoparametric hypersurface with g distinct principal curvatures $\cot(\theta_1 - t), \ldots, \cot(\theta_g - t)$. In that case, f_t is a parallel hypersurface of the hypersurface f.

If $\mu = \cot t$ is a principal curvature of f of multiplicity $m > 0$, then the map f_t is constant along each leaf of the m-dimensional principal foliation T_μ, and the image of f_t is a smooth submanifold of codimension $m + 1$ in S^n, called a *focal submanifold* of f. All of the hypersurfaces in a family of parallel isoparametric hypersurfaces have the same focal submanifolds, and Münzner showed that there are only two focal submanifolds, and at most two distinct multiplicities m_1, m_2 of the principal curvatures, regardless of the number g of distinct principal curvatures.

Münzner showed that if $f : M^{n-1} \to S^n$ is a connected isoparametric hypersurface with g distinct principal curvatures, then $f(M^{n-1})$ and its parallel hypersurfaces and focal submanifolds are each contained in a level set of a homogeneous polynomial F of degree g satisfying the *Cartan–Münzner differential equations* on the Euclidean differential operators grad F and Laplacian $\triangle F$ on $\mathbf{R}^{n+1}$,

$$| \operatorname{grad} F|^2 = g^2 r^{2g-2}, \quad r = |x|, \tag{4.134}$$

$$\triangle F = cr^{g-2}, \quad c = g^2(m_2 - m_1)/2,$$

where m_1, m_2 are the two (possibly equal) multiplicities of the principal curvatures on $f(M^{n-1})$.

Conversely, the level sets of the restriction $F|_{S^n}$ of such a function F to S^n constitute a family of parallel isoparametric hypersurfaces and their focal submanifolds, and F is called the *Cartan–Münzner polynomial* associated to this family of isoparametric hypersurfaces. Münzner also showed that these level sets are connected, and thus any connected isoparametric hypersurface in S^n lies in a unique compact connected isoparametric hypersurface obtained by taking the whole level set. The values of the restriction $F|_{S^n}$ range between -1 and $+1$, and the two focal submanifolds are

$$M_+ = (F|_{S^n})^{-1}(1), \qquad M_- = (F|_{S^n})^{-1}(-1).$$

Each principal curvature $\mu_i = \cot\theta_i$, $1 \le i \le g$, gives rise to two focal points corresponding to the values $t = \theta_i$ and $t = \theta_i + \pi$ in equation (4.133). The $2g$ focal points are evenly spaced along a normal geodesic to the family of parallel isoparametric hypersurfaces, and they lie alternately on the two focal submanifolds M_+ and M_-.

Now we begin the construction of Ferus, Karcher, and Münzner by recalling some facts about Clifford algebras. For each integer $m \ge 0$, the *Clifford algebra* C_m is the associative algebra over $\mathbf{R}$ that is generated by a unity 1 and the elements $e_1, \ldots, e_m$ subject only to the relations

$$e_i^2 = -1, \qquad e_i e_j = -e_j e_i, \quad i \ne j, \quad 1 \le i, j \le m. \tag{4.135}$$

One can show that the set

$$\{1, e_{i_1} \cdots e_{i_r} \mid i_1 < \cdots < i_r, \ 1 \le r \le m\}, \tag{4.136}$$

forms a basis for the underlying vector space C_m, and thus $\dim C_m = 2^m$.

Obviously, the Clifford algebra C_0 is isomorphic to $\mathbf{R}$, and C_1 is isomorphic to the field of complex numbers $\mathbf{C}$ with e_1 equal to the complex number i. Next one can show that C_2 is isomorphic to the skew-field of quaternions $\mathbf{H}$ with the correspondence

$$e_1 = i, \qquad e_2 = j, \qquad e_1 e_2 = ij = k.$$

In fact all of the Clifford algebras C_m have been explicitly determined by Atiyah, Bott and Shapiro [5] (see Table 4.1 below), and the classification of representations of Clifford algebras in [5] is crucial in the construction of Ferus, Karcher, and Münzner.

In order to explain the table below more fully, we need to introduce some terminology. We let $R(q)$ denote the algebra of $q \times q$ matrices with entries from an algebra R. The multiplication in $R(q)$ is defined by matrix multiplication using the operations of multiplication and addition defined in the algebra R. The *direct sum* $R_1 \oplus R_2$ is the Cartesian product $R_1 \times R_2$ with all algebra operations defined coordinatewise. An algebra homomorphism

$$\rho : C_m \to \mathbf{R}(q), \quad \text{with } \rho(1) = I \quad \text{(identity matrix)}$$

is called a *representation* of C_m on $\mathbf{R}^q$ or a representation of C_m of *degree q*. Two representations ρ and $\tilde{\rho}$ of C_m on $\mathbf{R}^q$ are said to be *equivalent* if there exists a matrix $A \in GL(q, \mathbf{R})$ such that

$$\tilde{\rho}(x) = A\rho(x)A^{-1},$$

for each $x \in C_m$.

Each representation of C_m on $\mathbf{R}^q$ corresponds to a set of matrices

$$E_1, \ldots, E_m$$

in $GL(q, \mathbf{R})$ such that

$$E_i^2 = -I, \qquad E_i E_j = -E_j E_i, \quad i \neq j, \quad 1 \leq i, j \leq m. \tag{4.137}$$

Furthermore, each representation is equivalent to a representation for which the E_i are skew-symmetric (see, for example, Ozeki–Takeuchi [143, Part I, pp. 543–548] or Conlon [54, pp. 349–352]). From now on, we will assume that the E_i are skew-symmetric.

Remark 4.30. Note that if the E_i in equation (4.137) are skew-symmetric, then they also must be orthogonal. This can be seen as follows. For any $v, w \in \mathbf{R}^q$, we denote the Euclidean inner product on $\mathbf{R}^q$ by $v \cdot w$. Then using equation (4.137) and the fact that E_i is skew-symmetric, we have

$$E_i v \cdot E_i w = v \cdot E_i^t E_i w = v \cdot (-E_i)E_i w = v \cdot Iw = v \cdot w, \tag{4.138}$$

and thus E_i is orthogonal.

Atiyah, Bott, and Shapiro determined all of the Clifford algebras according to Table 4.1 below. Moreover, they showed that the Clifford algebra C_{m-1} has an irreducible representation of degree q if and only if $q = \delta(m)$ as in the table.

m	C_{m-1}	$\delta(m)$	
1	$\mathbf{R}$	1	
2	$\mathbf{C}$	2	
3	$\mathbf{H}$	4	
4	$\mathbf{H} \oplus \mathbf{H}$	4	(4.139)
5	$\mathbf{H}(2)$	8	
6	$\mathbf{C}(4)$	8	
7	$\mathbf{R}(8)$	8	
8	$\mathbf{R}(8) \oplus \mathbf{R}(8)$	8	
$k+8$	$C_{k-1}(16)$	$16\delta(k)$	

Table 4.1. Clifford algebras and the degree of an irreducible representation.

One can obtain reducible representations of C_{m-1} on $\mathbf{R}^q$ for $q = k\delta(m)$, $k > 1$, by taking a direct sum of k irreducible representations of C_{m-1} on $\mathbf{R}^{\delta(m)}$.

Closely related to representations of Clifford algebras are the symmetric Clifford systems used by Ferus, Karcher, and Münzner. Let $\mathrm{Sym}_n(\mathbf{R})$ be the space of symmetric $n \times n$ matrices with real entries. On $\mathrm{Sym}_n(\mathbf{R})$, we have the inner product

$$\langle A, B \rangle = \mathrm{trace}(AB)/n. \tag{4.140}$$

Note that if $A = [a^i_j]$ and $B = [b^i_j]$ are symmetric $n \times n$ matrices, then

$$(AB)_{ii} = \sum_{k=1}^n a^i_k b^k_i = \sum_{k=1}^n a^i_k b^i_k, \tag{4.141}$$

and thus $\mathrm{trace}(AB)$ is just the usual Euclidean dot product on the space of matrices $\mathbf{R}(n)$ considered as $\mathbf{R}^{n^2}$. Dividing by n adjusts the metric so that $\langle I, I \rangle = 1$, where I is the $n \times n$ identity matrix.

Definition 4.31. Let l, m be positive integers.

(i) The $(m + 1)$-tuple $(P_0, \ldots, P_m)$ with $P_i \in \mathrm{Sym}_{2l}(\mathbf{R})$ is called a *(symmetric) Clifford system* on $\mathbf{R}^{2l}$ if we have

$$P_i^2 = I, \quad P_i P_j = -P_j P_i, \quad i \neq j, \quad 0 \leq i, j \leq m. \tag{4.142}$$

(ii) Let $(P_0, \ldots, P_m)$ and $(Q_0, \ldots, Q_m)$ be Clifford systems on $\mathbf{R}^{2l}$, respectively $\mathbf{R}^{2n}$, then $(P_0 \oplus Q_0, \ldots, P_m \oplus Q_m)$ is a Clifford system on $\mathbf{R}^{2(l+n)}$, called the *direct sum* of $(P_0, \ldots, P_m)$ and $(Q_0, \ldots, Q_m)$.

(iii) A Clifford system $(P_0, \ldots, P_m)$ on $\mathbf{R}^{2l}$ is called *irreducible* if it is not possible to write $\mathbf{R}^{2l}$ as a direct sum of two positive-dimensional subspaces that are invariant under all of the P_i.

Remark 4.32. The transformations $P_0, \ldots, P_m$ in a Clifford system must be orthogonal on $\mathbf{R}^{2l}$. To see this note that

$$P_i x \cdot P_i y = x \cdot P_i^2 y = x \cdot I y = x \cdot y, \tag{4.143}$$

for all $x, y \in \mathbf{R}^{2l}$.

Next we find a correspondence between Clifford systems on $\mathbf{R}^{2l}$ and representations of Clifford algebras on $\mathbf{R}^l$. First we begin with a representation of a Clifford algebra C_{m-1} on $\mathbf{R}^l$ and produce a Clifford system $(P_0, \ldots, P_m)$ on $\mathbf{R}^{2l}$.

Suppose that $E_1, \ldots, E_{m-1}$ are skew-symmetric matrices in $\mathbf{R}(l)$ that satisfy equation (4.137) and thus determine a representation of the Clifford algebra C_{m-1} on $\mathbf{R}^l$. We write $\mathbf{R}^{2l} = \mathbf{R}^l \oplus \mathbf{R}^l$ and define symmetric transformations on $\mathbf{R}^{2l}$ as follows:

$$P_0(u, v) = (u, -v), \qquad P_1(u, v) = (v, u), \tag{4.144}$$
$$P_{1+i}(u, v) = (E_i v, -E_i u), \qquad 1 \leq i \leq m - 1.$$

Lemma 4.33. *The transformations $P_0, \ldots, P_m$ in equation* (4.144) *form a Clifford system on* $\mathbf{R}^{2l}$.

Proof. The transformations P_0 and P_1 are clearly symmetric. We now check that P_{1+i} is symmetric for $1 \le i \le m-1$. Since the matrices E_i are skew-symmetric, we have

$$P_{1+i}(u, v) \cdot (x, y) = (E_i v, -E_i u) \cdot (x, y) = (E_i v \cdot x) - (E_i u \cdot y) \quad (4.145)$$
$$= -v \cdot E_i x + u \cdot E_i y = (u, v) \cdot (E_i y, -E_i x)$$
$$= (u, v) \cdot P_{1+i}(x, y).$$

Next we check that the P_i satisfy (i) in the definition of a Clifford system. Again it is clear that $P_0^2 = P_1^2 = I$. Then by equation (4.137), we have

$$P_{1+i}^2(u, v) = P_{1+i}(E_i v, -E_i u) = (-E_i^2 u, -E_i^2 v) = (u, v). \quad (4.146)$$

Similarly, it is a straightforward calculation to show that

$$P_i P_j = -P_j P_i \quad \text{for } i \ne j, \quad 0 \le i, j \le m,$$

and thus $(P_0, \ldots, P_m)$ is a Clifford system. $\qquad\square$

Conversely, suppose that $(P_0, \ldots, P_m)$ is a Clifford system on $\mathbf{R}^{2l}$. Note that since $P_i^2 = I$, the eigenvalues of P_i must be ± 1 for $0 \le i \le m$. In particular, for P_0 we denote these eigenspaces by

$$E_+(P_0) = \{x \in \mathbf{R}^{2l} \mid P_0 x = x\}, \quad (4.147)$$
$$E_-(P_0) = \{x \in \mathbf{R}^{2l} \mid P_0 x = -x\}.$$

We now show how $(P_0, \ldots, P_m)$ leads to a representation of C_{m-1} on $\mathbf{R}^l$ by the following two lemmas.

Lemma 4.34. *For $1 \le j \le m$, the transformation P_j interchanges the eigenspaces $E_+(P_0)$ and $E_-(P_0)$, and so both of these spaces have dimension l. As a consequence, trace $P_j = 0$, for $0 \le j \le m$.*

Proof. Suppose that $x \in E_+(P_0)$ so that $P_0 x = x$. Then $P_j(P_0 x) = P_j x$, but we also have $P_j(P_0 x) = -P_0 P_j x$. So $P_0 P_j x = -P_j x$, and $P_j x$ is in $E_-(P_0)$. Similarly, one can show that P_j maps $E_-(P_0)$ into $E_+(P_0)$. Since the orthogonal transformation P_j interchanges these two spaces, they must have the same dimension. Furthermore, $\mathbf{R}^{2l} = E_+(P_0) \oplus E_-(P_0)$, because P_0 is diagonalizable, and so the two spaces must both have dimension l. This shows that trace $P_0 = 0$, and a similar proof shows that trace $P_i = 0$ for all i. $\qquad\square$

We now have $\mathbf{R}^{2l} = E_+(P_0) \oplus E_-(P_0)$. The lemma implies $E_+(P_0)$ is invariant under $P_1 P_{1+i}$, for $1 \le i \le m-1$, since P_{1+i} maps $E_+(P_0)$ to $E_-(P_0)$, and then P_1 maps $E_-(P_0)$ to $E_+(P_0)$. We now identify $\mathbf{R}^l$ with $E_+(P_0)$ and define the maps $E_i : \mathbf{R}^l \to \mathbf{R}^l$ to be the restriction of $P_1 P_{1+i}$ to the invariant subspace $E_+(P_0)$, that is,

$$E_i = P_1 P_{1+i}|_{E_+(P_0)}, \quad 1 \le i \le m-1. \quad (4.148)$$

Lemma 4.35. *The transformations $E_1, \ldots, E_{m-1}$ in equation (4.148) are skew-symmetric, and they determine a representation of the Clifford algebra C_{m-1} on $\mathbf{R}^l$.*

Proof. We first show that the E_i are skew-symmetric. Let x and y be points in $\mathbf{R}^l = E_+(P_0)$. Then since P_1 and P_{1+i} are symmetric transformations on $\mathbf{R}^{2l} = E_+(P_0) \oplus E_-(P_0)$, we have from equation (4.142) that

$$P_1 P_{1+i}(x, 0) \cdot (y, 0) = (x, 0) \cdot P_{1+i} P_1(y, 0) = (x, 0) \cdot (-P_1 P_{1+i})(y, 0),$$

and thus $E_i x \cdot y = x \cdot (-E_i y)$, and E_i is skew-symmetric. Next we show that $E_i^2 = -I$. We have for $x \in \mathbf{R}^l$,

$$(P_1 P_{1+i})(P_1 P_{1+i})(x, 0) = P_1(-P_1 P_{1+i})P_{1+i}(x, 0) = (-I)(I)(x, 0) = (-x, 0),$$

so $E_i^2 x = -x$. Finally we show that $E_i E_j = -E_j E_i$ for $i \neq j$ as follows:

$$\begin{aligned}(P_1 P_{1+i})(P_1 P_{1+j})(x, 0) &= -(P_1 P_{1+i})(P_{1+j} P_1)(x, 0) \\ &= (P_1 P_{1+j})(P_{1+i} P_1)(x, 0) = -(P_1 P_{1+j})(P_1 P_{1+i})(x, 0),\end{aligned}$$

and so $E_i E_j x = -E_j E_i x$. Thus, $E_1, \ldots, E_{m-1}$ determine a representation of C_{m-1} on $\mathbf{R}^l$. $\qquad\square$

Given this correspondence between Clifford systems and representations of Clifford algebras, Ferus, Karcher, and Münzner deduce several facts about Clifford systems from known results in [5] concerning representations of Clifford algebras which we describe below.

First of all, a Clifford system is irreducible if and only if the corresponding Clifford algebra representation is irreducible, and thus an irreducible Clifford system $(P_0, \ldots, P_m)$ on $\mathbf{R}^{2l}$ exists precisely when $l = \delta(m)$ as in Table 4.1.

There is also a notion of equivalence of Clifford systems similar to the definition of equivalence of representations of Clifford algebras as follows.

Definition 4.36. Two Clifford systems $(P_0, \ldots, P_m)$ and $(Q_0, \ldots, Q_m)$ on $\mathbf{R}^{2l}$ are said to be *algebraically equivalent* if there exists an orthogonal transformation $A \in O(2l)$ such that $Q_i = A P_i A^t$, for $0 \leq i \leq m$. Two Clifford systems are said to be *geometrically equivalent* if there exists

$$B \in O(\mathrm{Span}\{P_0, \ldots, P_m\} \subset \mathrm{Sym}_{2l}(\mathbf{R}))$$

such that $(Q_0, \ldots, Q_m)$ and $(B P_0, \ldots, B P_m)$ are algebraically equivalent.

For $m \not\equiv 0 \pmod 4$, there exists exactly one algebraic equivalence class of irreducible Clifford systems. Thus, in this case there can be only one geometric equivalence class also. Hence, for each positive integer k there exists exactly one algebraic (or geometric) equivalence class of Clifford systems $(P_0, \ldots, P_m)$ on $\mathbf{R}^{2l}$ with $l = k\delta(m)$.

For $m \equiv 0 \pmod 4$, there exist exactly two algebraic classes of irreducible Clifford systems. These can be distinguished from each other by the choice of sign in

$$\text{trace}(P_0 \cdots P_m) = \pm \text{trace} \, I = \pm 2\delta(m). \tag{4.149}$$

There is only one geometric equivalence class of irreducible Clifford systems in this case also, as can be seen by replacing P_0 by $-P_0$. The absolute trace,

$$|\text{trace}(P_0 \cdots P_m)|, \tag{4.150}$$

is obviously an invariant under geometric equivalence. If one constructs all possible direct sums using both of the algebraic equivalence classes of irreducible Clifford systems with altogether k summands, then this invariant takes on $[k/2]+1$ different values, where $[k/2]$ is the greatest integer less than or equal to $k/2$. Thus, for $m \equiv 0$ (mod 4), there are exactly $[k/2]+1$ distinct geometric equivalence classes of Clifford systems on $\mathbf{R}^{2l}$ with $l = k\delta(m)$.

Definition 4.37. Let $(P_0, \ldots, P_m)$ be a Clifford system on $\mathbf{R}^{2l}$. The unit sphere in $\text{Span}\{P_0, \ldots, P_m\} \subset \text{Sym}_{2l}(\mathbf{R})$ is called the *Clifford sphere* determined by the system and is denoted $\Sigma(P_0, \ldots, P_m)$.

This is an important idea in this context, because the construction of Ferus, Karcher, and Münzner depends only on the Clifford sphere $\Sigma(P_0, \ldots, P_m)$ and not on the specific choice of $(P_0, \ldots, P_m)$. We therefore list several important properties of the Clifford sphere in the following lemma.

Lemma 4.38 (properties of the Clifford sphere). *The Clifford sphere $\Sigma(P_0, \ldots, P_m)$ has the following properties:*

(a) *For each $P \in \Sigma(P_0, \ldots, P_m)$, we have $P^2 = I$. Conversely, if Σ is the unit sphere in a linear subspace V spanned by Σ in $\text{Sym}_{2l}(\mathbf{R})$ such that $P^2 = I$ for all $P \in \Sigma$, then every orthonormal basis of V is a Clifford system on $\mathbf{R}^{2l}$.*

(b) *Two Clifford systems are geometrically equivalent if and only if their Clifford spheres are conjugate to one another under an orthogonal transformation of $\mathbf{R}^{2l}$.*

(c) *The function,*

$$H(x) = \sum_{i=0}^{m} (P_i x \cdot x)^2, \tag{4.151}$$

depends only on $\Sigma(P_0, \ldots, P_m)$ and not on the choice of orthonormal basis $(P_0, \ldots, P_m)$. For $P \in \Sigma(P_0, \ldots, P_m)$, we have

$$H(Px) = H(x), \tag{4.152}$$

for all x in $\mathbf{R}^{2l}$.

(d) *For an orthonormal set $\{Q_1, \ldots, Q_r\}$ in $\Sigma(P_0, \ldots, P_m)$, since $Q_i Q_j = -Q_j Q_i$, for $i \neq j$, we have*

$$Q_1 \cdots Q_r \text{ is symmetric if } r \equiv 0, 1 \mod 4, \tag{4.153}$$

$$Q_1 \cdots Q_r \text{ is skew-symmetric if } r \equiv 2, 3 \mod 4.$$

Furthermore, the product $Q_1 \cdots Q_r$ is uniquely determined by a choice of orientation of $\text{Span}\{Q_1, \ldots, Q_r\}$.

(e) *For* $P, Q \in \mathrm{Span}\{P_0, \ldots, P_m\}$ *and* $x \in \mathbf{R}^{2l}$, *we have*

$$Px \cdot Qx = \langle P, Q \rangle (x \cdot x). \tag{4.154}$$

Proof.

(a) Let $P = \sum_{i=0}^{m} a_i P_i$ with $\sum_{i=0}^{m} a_i^2 = 1$. Then

$$P^2 = \left(\sum_{i=0}^{m} a_i P_i \right) \left(\sum_{j=0}^{m} a_j P_j \right) = \sum_{i=0}^{m} \sum_{j=0}^{m} a_i a_j P_i P_j \tag{4.155}$$

$$= \sum_{i=0}^{m} a_i^2 P_i^2 + \sum_{i=0}^{m} \sum_{j \neq i} a_i a_j P_i P_j$$

$$= \sum_{i=0}^{m} a_i^2 I + \sum_{i=0}^{m} \sum_{j>i} a_i a_j (P_i P_j + P_j P_i) = \sum_{i=0}^{m} a_i^2 I = I.$$

Conversely, let $\{Q_0, \ldots, Q_m\}$ be an orthonormal basis for V. By hypothesis, $Q_i^2 = I$ for all i. We must show that $Q_i Q_j + Q_j Q_i = 0$ for all $i \neq j$. Let $Q = (1/\sqrt{2})(Q_i + Q_j)$, for $i \neq j$. Then Q has length 1, so $Q^2 = I$. On the other hand,

$$Q^2 = (Q_i^2 + (Q_i Q_j + Q_j Q_i) + Q_j^2)/2 \tag{4.156}$$

$$= (I + (Q_i Q_j + Q_j Q_i) + I)/2 = I + \frac{1}{2}(Q_i Q_j + Q_j Q_i),$$

and so $(Q_i Q_j + Q_j Q_i) = 0$.

(b) Two Clifford systems $(P_0, \ldots, P_m)$ and $(Q_0, \ldots, Q_m)$ on $\mathbf{R}^{2l}$ are geometrically equivalent if there exists an orthogonal transformation

$$B \in O(\mathrm{Span}\{P_0, \ldots, P_m\} \subset \mathrm{Sym}_{2l}(\mathbf{R}))$$

such that

$$Q_i = A(BP_i)A^t \quad \text{for } A \in O(2l). \tag{4.157}$$

Let $R_i = BP_i$. Then the Clifford spheres $\Sigma(P_0, \ldots, P_m)$ and $\Sigma(R_0, \ldots, R_m)$ are equal because B is orthogonal, and $\Sigma(Q_0, \ldots, Q_m)$ and $\Sigma(R_0, \ldots, R_m)$ are clearly conjugate by equation (4.157).

Conversely, if the Clifford spheres $\Sigma(P_0, \ldots, P_m)$ and $\Sigma(Q_0, \ldots, Q_m)$ are conjugate, then there exists $A \in O(2l)$ such that $\{Q_0, \ldots, Q_m\}$ is an orthonormal frame in the Clifford sphere

$$\Sigma(AP_0 A^t, \ldots, AP_m A^t) = \Sigma(Q_0, \ldots, Q_m).$$

So there must exist an orthogonal transformation $B \in O(\mathrm{Span}\{Q_0, \ldots, Q_m\})$ such that $BQ_i = AP_i A^t$, for $0 \leq i \leq m$, and so the Clifford systems $(P_0, \ldots, P_m)$ and $(Q_0, \ldots, Q_m)$ are geometrically equivalent.

(c) Suppose that $\{Q_0, \ldots, Q_m\}$ is another orthonormal basis for

$$\mathrm{Span}\{P_0, \ldots, P_m\}.$$

Then

$$Q_i = \sum_{j=0}^{m} b_i^j P_j, \quad [b_i^j] \in O(m+1).$$

Then we have

$$\sum_{i=0}^{m}(Q_ix \cdot x)^2 = \sum_{i=0}^{m}\left(\left(\sum_{j=0}^{m} b_i^j P_jx\right) \cdot x\right)^2 = \sum_{i=0}^{m}\left(\sum_{j=0}^{m} b_i^j (P_jx \cdot x)\right)^2. \quad (4.158)$$

Fix $x \in \mathbf{R}^{2l}$, and let $a_j = P_jx \cdot x$. Then the sum on the right side of equation (4.158) becomes

$$\sum_{i=0}^{m}(\sum_{j=0}^{m} b_i^j a_j)^2 = \sum_{i=0}^{m} a_i^2 = \sum_{i=0}^{m}(P_ix \cdot x)^2,$$

since $[b_i^j] \in O(m+1)$. So $H(x) = \sum_{i=0}^{m}(P_ix \cdot x)^2$ does not depend on the choice of orthonormal basis.

To show that $H(Px) = H(x)$ for $P \in \Sigma(P_0, \ldots, P_m)$, choose an orthonormal basis $\{Q_0, \ldots, Q_m\}$ for $\mathrm{Span}\{P_0, \ldots, P_m\}$ with $Q_0 = P$. Then

$$H(Px) = \sum_{i=0}^{m}(Q_i(Px) \cdot Px)^2 = \sum_{i=0}^{m}(Q_i Q_0x \cdot Q_0x)^2 \quad (4.159)$$

$$= (Q_0^2 x \cdot Q_0x)^2 + \sum_{i=1}^{m}(-Q_0 Q_ix \cdot Q_0x)^2$$

$$= (x \cdot Q_0x)^2 + \sum_{i=1}^{m}(Q_0 Q_ix \cdot Q_0x)^2$$

$$= (x \cdot Q_0x)^2 + \sum_{i=1}^{m}(Q_ix \cdot x)^2 = \sum_{i=0}^{m}(Q_ix \cdot x)^2 = H(x),$$

where we used the fact that Q_0 is orthogonal in going from the second to last line to the last line.

(d) For an orthonormal set $\{Q_1, \ldots, Q_r\}$ in $\Sigma(P_0, \ldots, P_m)$, since the Q_i are symmetric, we have

$$Q_1 \cdots Q_rx \cdot y = x \cdot Q_r \cdots Q_1y. \quad (4.160)$$

We use the equation $Q_i Q_j = -Q_j Q_i$ for $i \neq j$, to change $Q_r \cdots Q_1$ into $Q_1 \cdots Q_r$. The number of switches required is

$$(r-1) + (r-2) + \cdots + 1 = (r-1)r/2,$$

and this is even for $r \equiv 0, 1 \bmod 4$, and odd for $r \equiv 2, 3 \bmod 4$. Thus $Q_1 \cdots Q_r$ is symmetric for $r \equiv 0, 1 \bmod 4$, and skew-symmetric for $r \equiv 2, 3 \bmod 4$.

To see that $Q_1 \cdots Q_r$ is determined by an orientation of $\text{Span}\{Q_1, \ldots, Q_r\}$, note that $SO(r)$ is generated by rotations in two-dimensional coordinate planes. Since any two of the Q_i can be brought next to each other through interchanges using $Q_i Q_j = -Q_j Q_i$, it suffices to do the proof for $r = 2$, and this can be easily done by a direct calculation.

(e) First, it suffices to show the equation (4.154) for P_i and P_j in the orthonormal basis $\{P_0, \ldots, P_m\}$, since if

$$P = \sum_{i=0}^{m} a_i P_i, \qquad Q = \sum_{j=0}^{m} b_j P_j$$

we have using equation (4.154) for P_i and P_j,

$$Px \cdot Qx = \sum_{i=0}^{m} a_i P_i x \cdot \sum_{j=0}^{m} b_j P_j x = \sum_{i=0}^{m} \sum_{j=0}^{m} a_i b_j (P_i x \cdot P_j x)$$

$$= \sum_{i=0}^{m} \sum_{j=0}^{m} a_i b_j \langle P_i, P_j \rangle (x \cdot x) = \langle P, Q \rangle (x \cdot x),$$

since

$$\langle P, Q \rangle = \sum_{i=0}^{m} \sum_{j=0}^{m} a_i b_j \langle P_i, P_j \rangle.$$

Next we show that equation (4.154) holds for P_i and P_j. First, if $i = j$, since P_i is orthogonal, we have

$$P_i x \cdot P_i x = x \cdot x = 1(x \cdot x) = \langle P_i, P_i \rangle (x \cdot x).$$

Now suppose that $i \neq j$. Then $\langle P_i, P_j \rangle = 0$, so we must show that $P_i x \cdot P_j x = 0$ for all $x \in \mathbf{R}^{2l}$. Then $P_i x \cdot P_j x = x \cdot P_i P_j x$ and $P_i x \cdot P_j x = P_j P_i x \cdot x$, so

$$2(P_i x \cdot P_j x) = x \cdot (P_i P_j + P_j P_i)x = x \cdot 0 = 0,$$

as needed. $\qquad\qquad\qquad\qquad\qquad\qquad\qquad\qquad\qquad\qquad\qquad\qquad\qquad \square$

Now we can give the construction of Ferus, Karcher, and Münzner in the form of the following theorem. Note that by part (c) of Lemma 4.38, the function F below depends only on the Clifford sphere Σ and not on the choice of orthonormal basis $\{P_0, \ldots, P_m\}$.

Theorem 4.39. *Let $(P_0, \ldots, P_m)$ be a Clifford system on $\mathbf{R}^{2l}$. Let $m_1 = m$, $m_2 = l - m - 1$, and $F : \mathbf{R}^{2l} \to \mathbf{R}$ be defined by*

$$F(x) = (x \cdot x)^2 - 2 \sum_{i=0}^{m} (P_i x \cdot x)^2. \qquad (4.161)$$

Then F satisfies the Cartan–Münzner differential equations (4.134). If $m_2 > 0$, then the level sets of F on S^{2l-1} form a family of isoparametric hypersurfaces with $g = 4$ principal curvatures with multiplicities (m_1, m_2). (Note that the multiplicities satisfy $m_3 = m_1, m_4 = m_2$.)

Proof. By differentiating (4.161), we have

$$\text{grad } F = 4(x \cdot x)x - 8 \sum_{i=0}^{m} (P_i x \cdot x) P_i x. \tag{4.162}$$

Thus, we compute

$$|\text{grad } F|^2 = 16(x \cdot x)^3 - 64(x \cdot x) \sum_{i=0}^{m} (P_i x \cdot x)^2 \tag{4.163}$$

$$+ 64 \left(\sum_{i=0}^{m} (P_i x \cdot x) P_i x \cdot \sum_{j=0}^{m} (P_j x \cdot x) P_j x \right).$$

Then using equation (4.154) with $P = P_i$, $Q = P_j$, so that $\langle P_i, P_j \rangle = \delta_{ij}$, we have

$$\left(\sum_{i=0}^{m} (P_i x \cdot x) P_i x \cdot \sum_{j=0}^{m} (P_j x \cdot x) P_j x \right) = \sum_{i=0}^{m} \sum_{j=0}^{m} (P_i x \cdot x)(P_j x \cdot x)(P_i x \cdot P_j x)$$

$$= \sum_{i=0}^{m} \sum_{j=0}^{m} (P_i x \cdot x)(P_j x \cdot x)\langle P_i, P_j \rangle (x \cdot x)$$

$$= \sum_{i=0}^{m} (P_i x \cdot x)^2 (x \cdot x). \tag{4.164}$$

Substituting equation (4.164) into equation (4.163), we get

$$|\text{grad } F|^2 = 16(x \cdot x)^3 - 64(x \cdot x) \sum_{i=0}^{m} (P_i x \cdot x)^2 + 64(x \cdot x) \sum_{i=0}^{m} (P_i x \cdot x)^2$$

$$= 16(x \cdot x)^3 = 16r^6 = g^2 r^{2g-2}, \tag{4.165}$$

and thus we have the first differential equation in (4.134).

To show that we have the second differential equation in (4.134), we use the identity,

$$\Delta h^2 = 2|\text{grad } h|^2 + 2h\Delta h, \tag{4.166}$$

which holds for any smooth function $h : \mathbf{R}^{2l} \to \mathbf{R}$. We have

$$\Delta F = \Delta(x \cdot x)^2 - 2 \sum_{i=0}^{m} \Delta(P_i x \cdot x)^2. \tag{4.167}$$

We can use the identity in equation (4.166) on each term on the right side of equation (4.167). First, we take $h = x \cdot x$. Then grad $h = 2x$, and $\triangle h = 4l$, so

$$
\begin{aligned}
\triangle (x \cdot x)^2 &= 2|\operatorname{grad} h|^2 + 2h\triangle h \\
&= 8(x \cdot x) + 2(x \cdot x)4l \qquad\qquad (4.168) \\
&= 8(l + 1)(x \cdot x).
\end{aligned}
$$

Next we work with the term

$$
\sum_{i=0}^{m} \triangle (P_i x \cdot x)^2,
$$

in equation (4.167). Let $h_i = P_i x \cdot x$. Then by equation (4.166),

$$
\triangle h_i^2 = 2|\operatorname{grad} h_i|^2 + 2h_i\triangle h_i. \qquad\qquad (4.169)
$$

We compute grad $h_i = 2P_i x$, so

$$
|\operatorname{grad} h_i|^2 = 4(P_i x \cdot P_i x) = 4(x \cdot x). \qquad\qquad (4.170)
$$

Next one computes that

$$
\triangle h_i = \operatorname{trace} P_i = 0, \qquad\qquad (4.171)
$$

by Lemma 4.34. Thus the terms $h_i \triangle h_i$ in equation (4.169) are all zero. So we have from equations (4.169)–(4.171)

$$
\sum_{i=0}^{m} \triangle (P_i x \cdot x)^2 = \sum_{i=0}^{m} \triangle h_i^2 = \sum_{i=0}^{m} 8(x \cdot x) = 8(m + 1)(x \cdot x). \qquad\qquad (4.172)
$$

Combining equations (4.167), (4.168) and (4.172), we get

$$
\begin{aligned}
\triangle F &= 8(l + 1)(x \cdot x) - 16(m + 1)(x \cdot x) = 8((l - m - 1) - m)(x \cdot x) \\
&= 8(m_2 - m_1)(x \cdot x) = g^2 \left(\frac{m_2 - m_1}{2} \right) r^{g-2}, \qquad\qquad (4.173)
\end{aligned}
$$

so the second equation in (4.134) is satisfied.   □

As in the general theory of Münzner for isoparametric hypersurfaces, the function F in equation (4.161) takes values between -1 and $+1$ when restricted to the unit sphere S^{2l-1}. The two focal submanifolds are $M_+ = (F|_{S^n})^{-1}(1)$ and $M_- = (F|_{S^n})^{-1}(-1)$. We will concentrate on M_+ which turns out to be a Clifford–Stiefel manifold (see below). We can compute its principal curvatures, and thereby directly show that the hypersurfaces in the family, which are all tubes over M_+, are isoparametric. This gives a second proof of the fact that the level sets of the restriction of F to S^{2l-1} form a parallel family of isoparametric hypersurfaces and their focal submanifolds.

Since

$$F(x) = (x \cdot x)^2 - 2 \sum_{i=0}^{m} (P_i x \cdot x)^2,$$

we see that the subset M_+ of S^{2l-1} on which F takes the value 1 is precisely,

$$M_+ = \{x \in S^{2l-1} \mid P_i x \cdot x = 0, \quad 0 \le i \le m\}. \tag{4.174}$$

Recall the relationship between the orthogonal symmetric transformations P_i on $\mathbf{R}^{2l}$ and the orthogonal skew-symmetric transformations E_i on $\mathbf{R}^l$ given above in equation (4.144), that is,

$$P_0(u, v) = (u, -v), \quad P_1(u, v) = (v, u), \tag{4.175}$$
$$P_{1+i}(u, v) = (E_i v, -E_i u), \quad 1 \le i \le m - 1.$$

Thus we have for $(u, v) \in \mathbf{R}^l \times \mathbf{R}^l = \mathbf{R}^{2l}$,

$$P_0(u, v) \cdot (u, v) = |u|^2 - |v|^2, \quad P_1(u, v) \cdot (u, v) = 2(u \cdot v), \tag{4.176}$$
$$P_{1+i}(u, v) \cdot (u, v) = -2E_i u \cdot v, \quad 1 \le i \le m - 1.$$

Note that the equations

$$|u|^2 - |v|^2 = 0, \quad |u|^2 + |v|^2 = 1$$

imply that

$$|u|^2 = |v|^2 = 1/2. \tag{4.177}$$

Thus, we see that

$$M_+ = \{(u, v) \in S^{2l-1} \mid |u| = |v| = \frac{1}{\sqrt{2}}, u \cdot v = 0, E_i u \cdot v = 0, 1 \le i \le m - 1\}. \tag{4.178}$$

This is the so-called *Clifford–Stiefel manifold* $V_2(C_{m-1})$ of Clifford orthogonal 2-frames of length $1/\sqrt{2}$ in $\mathbf{R}^l$, where vectors u and v in $\mathbf{R}^l$ are said to be *Clifford orthogonal* if

$$u \cdot v = E_1 u \cdot v = \cdots = E_{m-1} u \cdot v = 0. \tag{4.179}$$

We now want to find the principal curvatures and the corresponding principal spaces for $M_+ = V_2(C_{m-1})$. We have

$$M_+ = \{x \in S^{2l-1} \mid P_i x \cdot x = 0, \quad 0 \le i \le m\}. \tag{4.180}$$

These are $m + 1$ independent conditions, so M_+ has codimension $m + 1$ in S^{2l-1}. At each point $x \in M_+$, the vectors $P_0 x, \dots, P_m x$ are all normal to M_+, since if X is tangent to M_+ at x, then we have

$$X(P_i x \cdot x) = 0,$$

but

$$X(P_i x \cdot x) = X(P_i x) \cdot x + P_i x \cdot X = P_i X \cdot x + P_i x \cdot X = 2(P_i x \cdot X),$$

since P_i is linear and symmetric. So $P_i x \cdot X = 0$ for all X tangent to M_+ at x, and $P_i x$ is normal to M_+ at x. Note also that

$$P_i x \cdot P_i x = x \cdot P_i^2 x = x \cdot x = 1, \tag{4.181}$$

and

$$P_i x \cdot P_j x = x \cdot P_i P_j x = -(x \cdot P_j P_i x) = -(P_j x \cdot P_i x), \tag{4.182}$$

so $P_i x \cdot P_j x = 0$ if $i \neq j$. We see that the set

$$\{P_0 x, \ldots, P_m x\} \tag{4.183}$$

is an orthonormal basis for the normal space $M_+^\perp(x)$ to M_+ in the sphere S^{2l-1}. Thus the normal bundle of M_+ is trivial with $\{P_0 x, \ldots, P_m x\}$ a global orthonormal frame. This also shows that

$$M_+^\perp(x) = \{Qx \mid Q \in \mathrm{Span}\{P_0, \ldots, P_m\}\}, \tag{4.184}$$

and the space of unit normals to M_+ at x is

$$B(x) = \{Px \mid P \in \Sigma(P_0, \ldots, P_m)\}. \tag{4.185}$$

We now want to determine the principal curvatures for the focal submanifold M_+. Let $\xi = Px$ be a unit normal to M_+ at a point x, where $P \in \Sigma(P_0, \ldots, P_m)$. We can extend ξ to a normal field on M_+ by setting $\xi(y) = Py$, for $y \in M_+$. Then the shape operator A_ξ is defined by

$$A_\xi X = -(\text{tangential component } \nabla_X \xi), \tag{4.186}$$

where X is a tangent vector to M_+ at a point x, and ∇ is the Riemannian covariant derivative on S^{2l-1}. Then using $\xi(y) = Py$, we compute

$$\nabla_X \xi = \nabla_X Py = P(X), \tag{4.187}$$

since P is a linear transformation on $\mathbf{R}^{2l}$, and so

$$A_\xi X = -(\text{tangential component } P(X)). \tag{4.188}$$

We can now compute the principal curvatures of A_ξ as follows.

Theorem 4.40. *Let x be a point on the focal submanifold M_+, and let $\xi = Px$ be a unit normal vector to M_+ at x, where $P \in \Sigma(P_0, \ldots, P_m)$. Let E_+ and E_- be the l-dimensional eigenspaces of P for the eigenvalues $+1$ and -1, respectively, so that $\mathbf{R}^{2l} = E_+ \oplus E_-$. Then the shape operator A_ξ has principal curvatures $0, 1, -1$ with corresponding principal spaces $T_0(\xi), T_1(\xi), T_{-1}(\xi)$ as follows:*

$$T_0(\xi) = \{QPx \mid Q \in \Sigma, \langle Q, P \rangle = 0\}, \tag{4.189}$$

$$T_1(\xi) = \{X \in E_- \mid X \cdot Qx = 0, \forall Q \in \Sigma\} = E_- \cap T_x M_+,$$
$$T_{-1}(\xi) = \{X \in E_+ \mid X \cdot Qx = 0, \forall Q \in \Sigma\} = E_+ \cap T_x M_+,$$

where $\Sigma = \Sigma(P_0, \ldots, P_m)$. *Furthermore,*

$$\dim T_0(\xi) = m, \quad \dim T_1(\xi) = \dim T_{-1}(\xi) = l - m - 1. \quad (4.190)$$

Proof. As we saw in equation (4.184), the normal space $M_+^\perp(x)$ to M_+ in S^{2l-1} is given by

$$M_+^\perp(x) = \{Qx \mid Q \in \mathrm{Span}\{P_0, \ldots, P_m\}\}.$$

This space has dimension $m + 1$, so

$$\dim M_+ = (2l - 1) - (m + 1) = 2l - m - 2 = m + 2(l - m - 1). \quad (4.191)$$

Suppose that $X = QPx$ for $Q \in \Sigma$, and $\langle Q, P \rangle = 0$. We first show that X is tangent to M_+ at x. We must show that X is orthogonal to every vector in $M_+^\perp(x)$. First, we have

$$X \cdot Px = QPx \cdot Px = -PQx \cdot Px = -Qx \cdot x = 0, \quad (4.192)$$
$$X \cdot Qx = QPx \cdot Qx = Px \cdot x = 0.$$

Next, suppose that $R \in \Sigma$ such that $\langle R, P \rangle = \langle R, Q \rangle = 0$. Then

$$X \cdot Rx = QPx \cdot Rx = RQPx \cdot x = -(x \cdot RQPx) = -(Rx \cdot QPx) = -(X \cdot Rx),$$

so $X \cdot Rx = 0$, where we have used the fact that RQP is skew-symmetric by part (d) of Lemma 4.38. Thus $X = QPx$ is in $T_x M_+$. Now we compute $A_\xi X$, for $\xi = Px$. By equation (4.187), we have

$$\nabla_X \xi = P(X) = P(QPx) = -(P^2 Qx) = -Qx,$$

which is normal to M_+ at x, so the tangential component $A_\xi X$ equals 0. Thus the m-dimensional space

$$\{QPx \mid Q \in \Sigma, \quad \langle Q, P \rangle = 0\} \subset T_0(\xi). \quad (4.193)$$

We will see that these sets are actually equal later.

Next, suppose that $X \in E_- \cap T_x M_+$. Then

$$\nabla_X \xi = P(X) = -X,$$

so $A_\xi X = X$, and $X \in T_1(\xi)$. So we have

$$E_- \cap T_x M_+ \subset T_1(\xi). \quad (4.194)$$

Note that since

$$E_- \cap T_x M_+ = \{X \in E_- \mid X \cdot Qx = 0, \forall Q \in \Sigma\},$$

this space has dimension $l - (m + 1) = l - m - 1$. Again, we will show later that the sets in equation (4.194) are equal.

Finally, suppose that $X \in E_+ \cap T_x M_+$. Then as above, we show that $A_\xi X = -X$, and so $X \in T_{-1}(\xi)$. Thus we have

$$E_+ \cap T_x M_+ \subset T_{-1}(\xi), \tag{4.195}$$

and this space also has dimension $l - m - 1$. Since the sum of the dimensions of the three mutually orthogonal spaces on the left sides of equations (4.193)–(4.195) is equal to $m + 2(l - m - 1) = \dim M_+$, the inclusions in equations (4.193)–(4.195) must all be equalities, and the theorem is proved. $\qquad\square$

Corollary 4.41. *Let M_t be a tube of spherical radius t over the focal submanifold M_+, where $0 < t < \pi$ and $t \notin \{\frac{\pi}{4}, \frac{\pi}{2}, \frac{3\pi}{4}\}$. Then M_t is an isoparametric hypersurface with four distinct principal curvatures,*

$$\cot(-t), \quad \cot\left(\frac{\pi}{4} - t\right), \quad \cot\left(\frac{\pi}{2} - t\right), \quad \cot\left(\frac{3\pi}{4} - t\right),$$

having respective multiplicities $m, l - m - 1, m, l - m - 1$.

Proof. This follows immediately from the formula for the principal curvatures of a tube (see, for example, [52, p. 132]), and the fact that M_+ has the principal curvatures and multiplicities given in Theorem 4.40. $\qquad\square$

From Theorem 4.39 or Corollary 4.41, we can determine the multiplicities of the principal curvatures of isoparametric hypersurfaces of FKM-type. The multiplicities are $m_1 = m$, which can be any positive integer, and $m_2 = l - m - 1$, where l is a positive integer such that the Clifford algebra C_{m-1} has a representation on $\mathbf{R}^l$. Thus, we know that $l = k\delta(m)$, where k is a positive integer and $\delta(m)$ is the unique positive integer such that C_{m-1} has an irreducible representation on $\mathbf{R}^{\delta(m)}$ (see Table 4.1). Thus, the multiplicities have the form

$$m_1 = m, \quad m_2 = k\delta(m) - m - 1, \quad k > 0. \tag{4.196}$$

Of course, the multiplicity m_2 must be positive in order for this construction to lead to an isoparametric hypersurface with four principal curvatures. In the table below, the cases where $m_2 \leq 0$ are denoted by a dash.

From parts (b) and (c) of Lemma 4.38 and from formula (4.161) for the Cartan–Münzner polynomial F, we see that geometrically equivalent Clifford systems determine congruent families of isoparametric hypersurfaces. In the table, the underlined multiplicities,

$$(\underline{m_1, m_2}), \quad (\underline{\underline{m_1, m_2}}),$$

denote the two, respectively, three geometrically inequivalent Clifford systems for the multiplicities (m_1, m_2). Ferus, Karcher, and Münzner show that these geometrically inequivalent Clifford systems with $m \equiv 0 \pmod 4$ and $l = k\delta(m)$ (as described earlier in this section) actually lead to incongruent families of isoparametric

$\delta(m)\|$	1	2	4	4	8	8	8	8	16	32	..
k											
1	–	–	–	–	(5, 2)	(6, 1)	–	–	(9, 6)	(10, 21)	..
2	–	(2, 1)	(3, 4)	(4, 3)	(5, 10)	(6, 9)	(7, 8)	(8, 7)	(9, 22)	(10, 53)	..
3	(1, 1)	(2, 3)	(3, 8)	(4, 7)	(5, 18)	(6, 17)	(7, 16)	(8, 15)	(9, 38)	(10, 85)	..
4	(1, 2)	(2, 5)	(3, 12)	(4, 11)	(5, 26)	(6, 25)	(7, 24)	(8, 23)	(9, 54)	·	..
5	(1, 3)	(2, 7)	(3, 16)	(4, 15)	(5, 34)	(6, 33)	(7, 32)	(8, 31)	·	·	..
·	·	·	·	·	·	·	·	·	·	·	..
·	·	·	·	·	·	·	·	·	·	·	..
·	·	·	·	·	·	·	·	·	·	·	..

Table 4.2. Multiplicities of principal curvatures of FKM-hypersurfaces.

hypersurfaces, of which there are $[k/2] + 1$. They also show that the families for multiplicities $(2, 1)$, $(6, 1)$, $(5, 2)$ and one of the $(4, 3)$-families are congruent to those with multiplicities $(1, 2)$, $(1, 6)$, $(2, 5)$ and $(3, 4)$, respectively, and these are the only coincidences under congruence among the FKM-hypersurfaces. The incongruence of families with the same multiplicities, as well as their inhomogeneity in many cases, is shown by Ferus, Karcher, and Münzner through a study of the second fundamental forms of the focal submanifolds.

As noted earlier, the construction of Ferus, Karcher, and Münzner gives all known examples of isoparametric hypersurfaces in a sphere with four principal curvatures with the exception of two homogeneous families having multiplicities $(2, 2)$ and $(4, 5)$. Stolz [177] proved that the multiplicities (m_1, m_2) of the principal curvatures of any isoparametric hypersurface with four principal curvatures must be the same as those of an FKM-hypersurface or one of the two homogeneous exceptions. For some time, it has been conjectured that the known examples are the only isoparametric hypersurfaces with four principal curvatures. Cecil, Chi and Jensen [40] have shown that if $M \subset S^n$ is an isoparametric hypersurface with four principal curvatures and $m_2 \geq 2m_1 - 1$, then M is of FKM-type (a different proof of this result, using isoparametric triple systems, was given by Immervoll [92]). This result, together with known classifications by Takagi [179] in the case $m_1 = 1$ and Ozeki and Takeuchi [143] for $m_1 = 2$, implies that there remain only four pairs of multiplicities, $(3, 4)$, $(6, 9)$, $(7, 8)$ and the homogeneous pair $(4, 5)$, for which the classification problem of isoparametric hypersurfaces with $g = 4$ principal curvatures remains an open question.

4.8 Compact Proper Dupin Submanifolds

In this section, we discuss compact proper Dupin submanifolds. The first examples are those that are Lie equivalent to an isoparametric hypersurface in the sphere S^n. Münzner [123] showed that the number g of distinct principal curvatures of an isoparametric hypersurface in S^n must be 1, 2, 3, 4 or 6. Thorbergsson [190] then showed that the same restriction holds for a compact proper Dupin hypersurface M^{n-1} embedded in S^n. He first proved that M^{n-1} must be taut in S^n. Using tautness, he then showed that M^{n-1} divides S^n into two ball bundles over the first focal submanifolds on either side of M^{n-1} in S^n. This topological data is all that is required for Münzner's restriction on g.

Compact proper Dupin hypersurfaces in S^n have been classified in the cases $g = 1, 2$ and 3. In each case, M^{n-1} must be Lie equivalent to an isoparametric hypersurface. The case $g = 1$ is simply the case of umbilic hypersurfaces, i.e., hyperspheres in S^n. In the case $g = 2$, Cecil and Ryan [47] showed that M^{n-1} must be a cyclide of Dupin (see Section 5.5), and thus it is Möbius equivalent to a standard product of spheres

$$S^k(r) \times S^{n-1-k}(s) \subset S^n(1) \subset \mathbf{R}^{n+1}, \quad r^2 + s^2 = 1.$$

In the case $g = 3$, Miyaoka [111] proved that M^{n-1} must be Lie equivalent to an isoparametric hypersurface (see also [41]). Earlier, Cartan [17] had shown that an isoparametric hypersurface with $g = 3$ principal curvatures is a tube over a standard embedding of a projective plane $\mathbf{FP}^2$, for $\mathbf{F} = \mathbf{R}, \mathbf{C}, \mathbf{H}$ (quaternions) or $\mathbf{O}$ (Cayley numbers), in S^4, S^7, S^{13} and S^{25}, respectively. (For $\mathbf{F} = \mathbf{R}$, a standard embedding is a spherical Veronese surface. See, for example, Cecil–Ryan [52, pp. 296–299].) These results led to the widely held conjecture (see Cecil–Ryan [52, p. 184]) that every compact connected proper Dupin hypersurface embedded in S^n is Lie equivalent to an isoparametric hypersurface. All attempts to verify this conjecture in the cases $g = 4$ and 6 were unsuccessful. Finally, in 1988, Pinkall and Thorbergsson [152] and Miyaoka and Ozawa [121] gave two different methods for producing counterexamples to the conjecture with $g = 4$ principal curvatures. The method of Miyaoka and Ozawa also yields counterexamples to the conjecture in the case $g = 6$.

Pinkall and Thorbergsson proved that their examples are not Lie equivalent to an isoparametric hypersurface by showing that the Lie curvature does not have the constant value $\Psi = 1/2$, as required for a submanifold that is Lie equivalent to an isoparametric hypersurface (see Section 4.5). Miyaoka and Ozawa showed that the Lie curvatures are not constant on their examples, and so these examples cannot be Lie equivalent to an isoparametric hypersurface. In this section, we will present both of these constructions.

The construction of Pinkall and Thorbergsson begins with an isoparametric hypersurface of FKM-type (see Section 4.7) with four principal curvatures, or rather with one of its focal submanifolds M_+ as in equation (4.178). Let us recall some details of that construction.

Start with a representation of the Clifford algebra C_{m-1} on $\mathbf{R}^l$ determined by $l \times l$ skew-symmetric matrices

$$E_1, \ldots, E_{m-1}$$

satisfying the equations,

$$E_i^2 = -I, \quad E_i E_j = -E_j E_i, \quad i \neq j, \quad 1 \leq i, j \leq m - 1. \tag{4.197}$$

Note that as in Remark 4.30 that the E_i must also be orthogonal.

As in Section 4.7, two vectors u and v in $\mathbf{R}^l$ are said to be *Clifford orthogonal* if

$$u \cdot v = E_1 u \cdot v = \cdots = E_{m-1} u \cdot v = 0. \tag{4.198}$$

Then we showed in equation (4.178) that the focal submanifold M_+ is given by

$$M_+ = \{(u, v) \in S^{2l-1} \mid |u| = |v| = \frac{1}{\sqrt{2}}, u \cdot v = 0, E_i u \cdot v = 0, 1 \le i \le m - 1\}, \tag{4.199}$$

where S^{2l-1} is the unit sphere in $\mathbf{R}^{2l} = \mathbf{R}^l \times \mathbf{R}^l$. This is the Clifford–Stiefel manifold $V_2(C_{m-1})$ of Clifford orthogonal 2-frames of length $1/\sqrt{2}$ in $\mathbf{R}^l$. Note that $M_+ = V_2(C_{m-1})$ is a submanifold of codimension $m + 1$ in S^{2l-1}.

In Theorem 4.40, we showed that for any unit normal ξ at any point $(u, v) \in V_2(C_{m-1})$, the shape operator A_ξ has three distinct principal curvatures

$$\kappa_1 = -1, \quad \kappa_2 = 0, \quad \kappa_3 = 1, \tag{4.200}$$

with respective multiplicities $l - m - 1, m, l - m - 1$.

The submanifold $V_2(C_{m-1})$ of codimension $m + 1$ in S^{2l-1} induces a Legendre submanifold defined on the unit normal bundle $B(V_2(C_{m-1}))$ of $V_2(C_{m-1})$ in S^{2l-1}. As in Theorem 4.15, this Legendre submanifold has a fourth principal curvature $\kappa_4 = \infty$ of multiplicity m at each point of $B(V_2(C_{m-1}))$. Since $\kappa_4 = \infty$, the Lie curvature Ψ at any point of $B(V_2(C_{m-1}))$ equals the Möbius curvature Φ, i.e.,

$$\Psi = \Phi = \frac{\kappa_1 - \kappa_2}{\kappa_1 - \kappa_3} = \frac{-1 - 0}{-1 - 1} = \frac{1}{2}. \tag{4.201}$$

Since all four principal curvatures are constant on $B(V_2(C_{m-1}))$, a tube M_t of spherical radius t, where $0 < t < \pi$ and $t \notin \{\frac{\pi}{4}, \frac{\pi}{2}, \frac{3\pi}{4}\}$, over $V_2(C_{m-1})$ is an isoparametric hypersurface with four distinct principal curvatures, as in Corollary 4.41. Note that Münzner [123] proved that if M is any isoparametric hypersurface in S^n with four principal curvatures, then the Lie curvature $\Psi = 1/2$ on all of M.

We now begin the construction of Pinkall and Thorbergsson. Given positive real numbers α and β with

$$\alpha^2 + \beta^2 = 1, \quad \alpha \neq \frac{1}{\sqrt{2}}, \quad \beta \neq \frac{1}{\sqrt{2}},$$

let

$$T_{\alpha,\beta} : \mathbf{R}^{2l} \to \mathbf{R}^{2l},$$

be the linear map defined by

$$T_{\alpha,\beta}(u, v) = \sqrt{2}\,(\alpha u, \beta v).$$

Then for $(u, v) \in V_2(C_{m-1})$, we have

$$|T_{\alpha,\beta}(u, v)|^2 = 2(\alpha^2(u \cdot u) + \beta^2(v \cdot v)) = 2\left(\frac{\alpha^2}{2} + \frac{\beta^2}{2}\right) = 1,$$

and thus the image $V_2^{\alpha,\beta} = T_{\alpha,\beta}V_2(C_{m-1})$ is a submanifold of S^{2l-1} of codimension $m + 1$ also.

Furthermore, we can see that the Legendre submanifold induced by $V_2^{\alpha,\beta}$ is proper Dupin as follows. (For the sake of brevity, we will say that a submanifold $V \subset S^n$ of codimension greater than one is proper Dupin if its induced Legendre submanifold is proper Dupin.)

We use the notion of curvature surfaces of a submanifold of codimension greater than one defined by Reckziegel [158] (see Remark 4.16). Specifically, suppose that $V \subset S^n$ is a submanifold of codimension greater than one, and let $B(V)$ denote its unit normal bundle. A connected submanifold $S \subset V$ is called a curvature surface if there exits a parallel section $\eta : S \to B(V)$ such that for each $x \in S$, the tangent space $T_x S$ is equal to some eigenspace of $A_{\eta(x)}$. The corresponding principal curvature $\kappa : S \to \mathbf{R}$ is then a smooth function on S. Reckziegel showed that if a principal curvature κ has constant multiplicity μ on $B(V)$ and is constant along each of its curvature surfaces, then each of its curvature surfaces is an open subset of a μ-dimensional metric sphere in S^n. Since our particular submanifold $V_2(C_{m-1})$ is compact, all of the curvature surfaces of the principal curvatures κ_1, κ_2 and κ_3 given in equation (4.200) are spheres of the appropriate dimensions in S^{2l-1}.

We now show that the Legendre submanifold induced by $V_2^{\alpha,\beta}$ is proper Dupin with four principal curvatures

$$\lambda_1 < \lambda_2 < \lambda_3 < \lambda_4.$$

Since $V_2^{\alpha,\beta}$ has codimension $m + 1$, the principal curvature $\lambda_4 = \infty$ has multiplicity m and is constant along its curvature surfaces. To complete the proof that $V_2^{\alpha,\beta}$ is proper Dupin, we establish a bijective correspondence between the other curvature surfaces of $V_2(C_{m-1})$ and those of $V_2^{\alpha,\beta}$. Let S be any curvature surface of $V_2(C_{m-1})$. Since $V_2(C_{m-1})$ is compact and proper Dupin, S is a μ-dimensional sphere, where μ is the multiplicity of the corresponding principal curvature of $V_2(C_{m-1})$. Along the curvature surface S, the corresponding curvature sphere Σ is constant. Note that Σ is a hypersphere obtained by intersecting S^{2l-1} with a hyperplane π that is tangent to $V_2(C_{m-1})$ along S. The image $T_{\alpha,\beta}(\pi)$ is a hyperplane that is tangent to $V_2^{\alpha,\beta}$ along the μ-dimensional sphere $T_{\alpha,\beta}(S)$. Since the hypersphere $T_{\alpha,\beta}(\pi) \cap S^{2l-1}$ is tangent to $V_2^{\alpha,\beta}$ along $T_{\alpha,\beta}(S)$, it is a curvature sphere of $V_2^{\alpha,\beta}$ with multiplicity μ, and $T_{\alpha,\beta}(S)$ is the corresponding curvature surface. Thus, we have a bijective correspondence between the curvature surfaces of $V_2(C_{m-1})$ and those of $V_2^{\alpha,\beta}$, and the Dupin condition is clearly satisfied on $V_2^{\alpha,\beta}$. Therefore, $V_2^{\alpha,\beta}$ is a proper Dupin submanifold with four principal curvatures, including $\lambda_4 = \infty$.

We next show that the Legendre submanifold induced by $V_2^{\alpha,\beta}$ is not Lie equivalent to an isoparametric hypersurface in S^{2l-1} by showing that the Lie curvature Ψ does not equal $1/2$ at some points of the unit normal bundle $B(V_2^{\alpha,\beta})$. First, note that

$$V_2^{\alpha,\beta} \subset f^{-1}(0) \cap g^{-1}(0),$$

where f and g are the real-valued functions defined on S^{2l-1} by

$$f(u, v) = \frac{-\beta}{2\alpha} u \cdot u + \frac{\alpha}{2\beta} v \cdot v, \qquad g(u, v) = -u \cdot v.$$

Thus, the gradients,

$$\xi = \left(\frac{-\beta}{\alpha} u, \frac{\alpha}{\beta} v\right), \qquad \eta = (-v, -u),$$

of f and g are two unit normal vector fields on $V_2^{\alpha,\beta}$. Note that by Theorem 4.39, we have $l > m + 1$ for the FKM-hypersurfaces, so we can choose $x, y \in \mathbf{R}^l$ such that

$$|x| = \alpha, \quad x \cdot u = 0, \quad x \cdot v = 0, \quad x \cdot E_i v = 0, \quad 1 \le i \le m - 1, \qquad (4.202)$$
$$|y| = \beta, \quad y \cdot u = 0, \quad y \cdot v = 0, \quad y \cdot E_i u = 0, \quad 1 \le i \le m - 1.$$

We define three curves,

$$\gamma(t) = (\cos t\, u + \sin t\, x, v), \qquad \delta(t) = (u, \cos t\, v + \sin t\, y), \qquad (4.203)$$
$$\varepsilon(t) = (\cos t\, u + \frac{\alpha}{\beta} \sin t\, v, -\frac{\beta}{\alpha} \sin t\, u + \cos t\, v).$$

It is straightforward to check that each of these curves lies on $V_2^{\alpha,\beta}$ and goes through the point (u, v) when $t = 0$. Along γ, the normal vector ξ is given by

$$\xi(t) = \left(-\frac{\beta}{\alpha}(\cos t\, u + \sin t\, x), \frac{\alpha}{\beta} v\right).$$

Thus,

$$\xi'(0) = \left(-\frac{\beta}{\alpha} x, 0\right) = -\frac{\beta}{\alpha} \gamma'(0).$$

So $X = (x, 0) = \gamma'(0)$ is a principal vector of A_ξ at (u, v) with corresponding principal curvature β/α.

Similarly, $Y = (0, y) = \delta'(0)$ is a principal vector of A_ξ at (u, v) with corresponding principal curvature $-\alpha/\beta$. Finally, along the curve ε, we have

$$\xi(t) = \left(-\frac{\beta}{\alpha}\left(\cos t\, u + \frac{\alpha}{\beta} \sin t\, v\right), \frac{\alpha}{\beta}\left(-\frac{\beta}{\alpha} \sin t\, u + \cos t\, v\right)\right).$$

Then $\xi'(0) = (-v, -u) = \eta$, which is normal to $V_2^{\alpha,\beta}$ at (u, v). Thus, we have $A_\xi Z = 0$, for $Z = \varepsilon'(0)$, and Z is a principal vector with corresponding principal curvature zero. Therefore, at the point $\xi(u, v)$ in $B(V_2^{\alpha,\beta})$, there are four principal curvatures,

$$\lambda_1 = -\frac{\alpha}{\beta}, \qquad \lambda_2 = 0, \qquad \lambda_3 = \frac{\beta}{\alpha}, \qquad \lambda_4 = \infty.$$

At this point, the Lie curvature Ψ is

$$\Psi = \Phi = \frac{\lambda_1 - \lambda_2}{\lambda_1 - \lambda_3} = \frac{-\alpha/\beta}{(-\alpha/\beta - \beta/\alpha)} = \alpha^2.$$

Since $\alpha^2 \neq 1/2$, the Legendre lift of $V_2^{\alpha,\beta}$ is not Lie equivalent to an isoparametric hypersurface. To obtain a compact proper Dupin hypersurface in S^{2l-1} with four principal curvatures that is not Lie equivalent to an isoparametric hypersurface, one simply takes a tube M over $V_2^{\alpha,\beta}$ in S^{2l-1} of sufficiently small radius so that the tube is an embedded hypersurface.

We remark that the Lie curvature is not constant on M. This follows from a theorem of Miyaoka [113, Corollary 8.3, p. 252] which states that if the Lie curvature Ψ is constant on a compact, connected proper Dupin hypersurface with four principal curvatures, then, in fact, $\Psi = 1/2$ on the hypersurface.

Next we handle the construction of the examples due to Miyaoka and Ozawa [121]. The key ingredient here is the Hopf fibration of S^7 over S^4. Let $\mathbf{R}^8 = \mathbf{H} \times \mathbf{H}$, where $\mathbf{H}$ is the skew field of quaternions. The Hopf fibering of the unit sphere S^7 in $\mathbf{R}^8$ over S^4 is given by

$$h(u, v) = (2u\bar{v}, |u|^2 - |v|^2), \quad u, v \in \mathbf{H}. \tag{4.204}$$

One can easily compute that the image of h lies in the unit sphere S^4 in the Euclidean space $\mathbf{R}^5 = \mathbf{H} \times \mathbf{R}$.

Before beginning the construction of Miyaoka and Ozawa, we recall some facts about the Hopf fibration. Suppose $(w, t) \in S^4$, with $t \neq 1$, i.e., (w, t) is not the point $(0, 1)$. We want to find the inverse image of (w, t) under h. Suppose that

$$2u\bar{v} = w, \quad |u|^2 - |v|^2 = t. \tag{4.205}$$

Multiplying the first equation in (4.205) by v on the right, we obtain

$$2u|v|^2 = wv, \quad 2u = \frac{w}{|v|}\frac{v}{|v|}. \tag{4.206}$$

Since $|u|^2 + |v|^2 = 1$, the second equation in (4.205) yields

$$|v|^2 = (1 - t)/2. \tag{4.207}$$

If we write $z = v/|v|$, then $z \in S^3$, the unit sphere in $\mathbf{H} = \mathbf{R}^4$. Then equations (4.206) and (4.207) give

$$u = \frac{wz}{\sqrt{2(1-t)}}, \quad v = \sqrt{(1-t)/2}\, z, \quad z \in S^3. \tag{4.208}$$

Thus, if U is the open set $S^4 - \{(0, 1)\}$, then $h^{-1}(U)$ is diffeomorphic to $U \times S^3$ by the formula (4.208). Of course, the second equation in (4.205) shows that $h^{-1}(0, 1)$ is just the 3-sphere in S^7 determined by the equation $v = 0$. We can find a similar local trivialization containing these points with $v = 0$ by beginning the process above with multiplication of equation (4.205) by $\bar{u}$ on the left, rather than by v on the right. As a consequence of this local triviality, if M is an embedded submanifold in S^4 which does not equal all of S^4, then $h^{-1}(M)$ is diffeomorphic to $M \times S^3$. Finally, recall that the Euclidean inner product on the space $\mathbf{R}^8 = \mathbf{H} \times \mathbf{H}$ is given by

$$(a, b) \cdot (u, v) = \Re(\bar{a}u + \bar{b}v), \tag{4.209}$$

where $\Re w$ denotes the real part of the quaternion w.

The examples of Miyaoka and Ozawa all arise as inverse images under h of proper Dupin hypersurfaces in S^4. The proof that these examples are proper Dupin is accomplished by first showing that they are taut. Thus, we begin with the following.

Theorem 4.42. *Let M be a compact, connected submanifold of S^4. If M is taut in S^4, then $h^{-1}(M)$ is taut in S^7.*

Proof. Since both M and $h^{-1}(M)$ lie in spheres, tautness of $h^{-1}(M)$ in S^7 is equivalent to tightness of $h^{-1}(M)$ in $\mathbf{R}^8$, i.e., every nondegenerate linear height function in $\mathbf{R}^8$ has the minimum number of critical points. We write linear height functions in $\mathbf{R}^8$ in the form

$$f_{ab}(u, v) = \Re(au + bv) = (\bar{a}, \bar{b}) \cdot (u, v), \quad (a, b) \in S^7. \tag{4.210}$$

This is the height function in the direction $(\bar{a}, \bar{b})$. We want to determine when the point (u, v) is a critical point of f_{ab}. Without loss of generality, we may assume that (u, v) lies in a local trivialization of the form (4.208) when making local calculations. Let $x = (w, t)$ be a point of $M \subset S^4$, and let

$$(x, z) = (w, t, z)$$

be a point in the fiber $h^{-1}(x)$. The tangent space to $h^{-1}(M)$ at (x, z) can be decomposed as $T_x M \times T_z S^3$. We first locate the critical points of the restriction of f_{ab} to the fiber through (x, z). By equations (4.208) and (4.210), we have

$$f_{ab}(w, t, z) = \Re \left(\frac{awz}{\sqrt{2(1-t)}} + bz\sqrt{(1-t)/2} \right) \tag{4.211}$$

$$= \Re(\alpha(w, t)z) = \alpha(w, t) \cdot \bar{z},$$

where

$$\alpha(w, t) = \frac{aw}{\sqrt{2(1-t)}} + b\sqrt{(1-t)/2}.$$

This defines the map α from S^4 to $\mathbf{H}$. If Z is any tangent vector to S^3 at z, we write Zf_{ab} for the derivative of f_{ab} in the direction $(0, Z)$. Then

$$Zf_{ab} = \alpha(w, t) \cdot \bar{Z} \tag{4.212}$$

at (x, z). Now there are two cases to consider. First, if $\alpha(w, t) \neq 0$, then in order to have $Zf_{ab} = 0$ for all $Z \in T_z S^3$, we must have

$$\bar{z} = \pm \frac{\alpha(w, t)}{|\alpha(w, t)|}. \tag{4.213}$$

So the restriction of f_{ab} to the fiber has exactly two critical points with corresponding values

$$\pm|\alpha(w, t)|. \tag{4.214}$$

The second case is when $\alpha(w, t) = 0$. Then the restriction of f_{ab} to the fiber is identically zero by equation (4.211). In both cases the function,

$$g_{ab}(w, t) = |\alpha(w, t)|^2,$$

satisfies the equation

$$g_{ab}(w, t) = f_{ab}^2(w, t, z),$$

at the critical point. The key in relating this fact to information about the submanifold M is to note that

$$\begin{aligned}
g_{ab}(w, t) = |\alpha(w, t)|^2 &= \frac{1}{2}\Re\{2a\bar{b}w + (|a|^2 - |b|^2)t\} + \frac{1}{2}(|a|^2 + |b|^2) \\
&= \frac{1}{2} + \frac{1}{2}((w, t) \cdot (2\bar{a}b, |a|^2 - |b|^2)) \tag{4.215} \\
&= \frac{1}{2} + \frac{1}{2}\ell_{ab}(w, t),
\end{aligned}$$

where ℓ_{ab} is the linear height function on $\mathbf{R}^5$ in the direction

$$(2\bar{a}b, |a|^2 - |b|^2) = h(\bar{a}, \bar{b}).$$

This shows that $g_{ab}(w, t) = 0$ if and only if $(w, t) = -h(\bar{a}, \bar{b})$. Thus, if $-h(\bar{a}, \bar{b})$ is not in M, the restriction of f_{ab} to each fiber has exactly two critical points of the form (x, z), with z as in equation (4.213). For $X \in T_x M$, we write $X f_{ab}$ for the derivative of f_{ab} in the direction $(X, 0)$. At the two critical points, we have

$$X f_{ab} = d\alpha(X) \cdot \bar{z}, \tag{4.216}$$
$$X g_{ab} = 2d\alpha(X) \cdot \alpha(x) = \pm 2|\alpha(x)|(d\alpha(X) \cdot \bar{z}) = \pm 2|\alpha(X)|X f_{ab}. \tag{4.217}$$

Thus (x, z) is a critical point of f_{ab} if and only if x is a critical point of g_{ab}. By equation (4.215), this happens precisely when x is a critical point of ℓ_{ab}. We conclude that if $-h(\bar{a}, \bar{b})$ is not in M, then f_{ab} has two critical points for every critical point of ℓ_{ab} on M. The set of points (a, b) in S^7 such that $-h(\bar{a}, \bar{b})$ belongs to M has measure zero. If (a, b) is not in this set, then f_{ab} has twice as many critical points as the height function ℓ_{ab} on M. Since M is taut, every nondegenerate height function ℓ_{ab} has $\beta(M; \mathbf{Z}_2)$ critical points on M, where $\beta(M; \mathbf{Z}_2)$ is the sum of the $\mathbf{Z}_2$-Betti numbers of M. Thus, except for a set of measure zero, every height function f_{ab} has $2\beta(M; \mathbf{Z}_2)$ critical points on $h^{-1}(M)$. Since $h^{-1}(M)$ is diffeomorphic to $M \times S^3$, we have

$$\beta(h^{-1}(M); \mathbf{Z}_2) = \beta(M \times S^3; \mathbf{Z}_2) = 2\beta(M; \mathbf{Z}_2).$$

Thus, $h^{-1}(M)$ is taut in S^7. $\square$

We next use Theorem 4.42 to show that the inverse image under h of a compact proper Dupin submanifold in S^4 is proper Dupin. Recall that a submanifold M of codimension greater than one is proper Dupin if the Legendre submanifold induced by M is proper Dupin. A taut submanifold is always Dupin, but it may not be proper Dupin (Pinkall [151], and Miyaoka [112], independently, for hypersurfaces), i.e., the number of distinct principal curvatures may not be constant on the unit normal bundle $B(M)$. Ozawa [142] proved that a taut submanifold $M \subset S^n$ is proper Dupin if and only if every connected component of a critical set of a linear height function on M is a point or is homeomorphic to a sphere of some dimension k. (See also Hebda [86].) This result is a key fact in the proof of the following theorem.

Theorem 4.43. *Let M be a compact, connected proper Dupin submanifold embedded in S^4. Then $h^{-1}(M)$ is a proper Dupin submanifold in S^7.*

Proof. As we have noted before, Thorbergsson [190] proved that a compact proper Dupin hypersurface embedded in S^n is taut, and Pinkall [151] extended this result to the case where M has codimension greater than one and the number of distinct principal curvatures is constant on the unit normal bundle $B(M)$. Thus, our M is taut in S^4, and therefore $h^{-1}(M)$ is taut in S^7 by Theorem 4.42. To complete the proof of the theorem, we need to show that each connected component of a critical set of a height function f_{ab} on $h^{-1}(M)$ is a point or a sphere.

We use the same notation as in the proof of Theorem 4.42. Now suppose that (x, z) is a critical point of f_{ab}. For $X \in T_x M$, we compute from equation (4.211) that

$$X f_{ab} = d\alpha(X) \cdot \bar{z}. \tag{4.218}$$

From (4.217), we see that $X g_{ab}$ also equals zero, and the argument again splits into two cases, depending on whether or not $g_{ab}(x)$ is zero. If $g_{ab}(x)$ is nonzero, then there are two critical points of f_{ab} on the fiber $h^{-1}(x)$. Thus a component in $h^{-1}(M)$ of the critical set of f_{ab} through (x, z) is homeomorphic to the corresponding component of the critical set containing x of the linear function ℓ_{ab} on M. Since M is proper Dupin, such a component is a point or a sphere.

The second case is when $g_{ab}(x) = f_{ab}^2(x, z) = 0$. As we have seen, this happens only if $x = -h(\bar{a}, \bar{b})$. In that case, x is an isolated absolute minimum of the height function ℓ_{ab}. Thus, the corresponding component of the critical set of f_{ab} through (x, z) lies in the fiber $h^{-1}(x)$, which is diffeomorphic to S^3. From equation (4.218), we see that this component of the critical set consists of those points (x, y) in the fiber such that $\bar{y}$ is orthogonal to $d\alpha(X)$, for all $X \in T_x M$. We know that

$$g_{ab}(x) = \frac{1}{2} + \frac{1}{2}\ell_{ab}(x), \tag{4.219}$$

and x is an isolated critical point of ℓ_{ab} on M. The tautness of M and the results of Ozawa [142] imply that x is a nondegenerate critical point of ℓ_{ab}, since the component of the critical set of a height function containing a degenerate critical point must be a sphere of dimension greater than zero. By equation (4.219), x is also a nondegenerate critical point of g_{ab}, and so the Hessian $H(X, Y)$ of g_{ab} is nondegenerate at x. Since $\alpha(x) = 0$, we compute that for X and Y in $T_x M$,

$$H(X, Y) = 2d\alpha(X) \cdot d\alpha(Y).$$

Hence, $d\alpha$ is nondegenerate at x, and the rank of $d\alpha$ is the dimension of M. From this it follows that the component of the critical set of f_{ab} through (x, z) is a sphere in $h^{-1}(x)$ of dimension $(3 - \dim M)$. Therefore, we have shown that every component of the critical set of a linear height function f_{ab} on $h^{-1}(M)$ is homeomorphic to a point or a sphere. Thus, $h^{-1}(M)$ is proper Dupin. □

Next, we relate the principal curvatures of $h^{-1}(M)$ to those of M.

Theorem 4.44. *Let M be a compact, connected proper Dupin hypersurface embedded in S^4 with g principal curvatures. Then the proper Dupin hypersurface $h^{-1}(M)$ in S^7 has 2g principal curvatures. Each principal curvature,*

$$\lambda = \cot \theta, \quad 0 < \theta < \pi,$$

of M at a point $x \in M$ yields two principal curvatures of $h^{-1}(M)$ at points in $h^{-1}(x)$ with values

$$\lambda^+ = \cot(\theta/2), \quad \lambda^- = \cot((\theta + \pi)/2).$$

Proof. A principal curvature $\lambda = \cot \theta$ of a hypersurface M at x corresponds to a focal point at oriented distance θ along the normal geodesic to M at x. (See, for example, Cecil–Ryan [52, p. 127].) A point (x, z) in $h^{-1}(M)$ is a critical point of f_{ab} if and only if $(\bar{a}, \bar{b})$ lies along the normal geodesic to $h^{-1}(M)$ at (x, z). The critical point is degenerate if and only if $(\bar{a}, \bar{b})$ is a focal point of $h^{-1}(M)$ at (x, z). Note further that (x, z) is a degenerate critical point of f_{ab} if and only if x is a degenerate critical point of ℓ_{ab}. This follows from the fact that both embeddings are taut, and the dimensions of the components of the critical sets agree by Theorem 4.43. The latter claim holds even when $x = -h(\bar{a}, \bar{b})$, since the fact that M has dimension three implies that the critical point (x, z) of f_{ab} is isolated. Thus, $(\bar{a}, \bar{b})$ is a focal point of $h^{-1}(M)$ if and only if $h(\bar{a}, \bar{b})$ is a focal point of M. Suppose now that $(\bar{a}, \bar{b})$ lies along the normal geodesic to $h^{-1}(M)$ at (x, z) and that $f_{ab}(x, z) = \cos \phi$. Then by equation (4.215),

$$g_{ab}(x) = \frac{1}{2} + \frac{1}{2}\ell_{ab}(x) = \frac{1}{2} + \frac{1}{2}\cos \theta,$$

where θ is the distance from $h(\bar{a}, \bar{b})$ to x. Since (x, z) is a critical point of f_{ab}, we have $g_{ab}(x) = f_{ab}^2(x, z)$. Thus,

$$\frac{1}{2} + \frac{1}{2}\cos \theta = \cos^2 \phi = \frac{1}{2} + \frac{1}{2}\cos 2\phi,$$

and so $\cos \theta = \cos 2\phi$. This means that under the map h, the normal geodesic to $h^{-1}(M)$ at (x, z) double covers the normal geodesic to M at x, since the points corresponding to the values $\phi = \theta/2$ and $\phi = (\theta + \pi)/2$ are mapped to the same point by h. In particular, a focal point corresponding to a principal curvature $\lambda = \cot \theta$ on the normal geodesic to M at x gives rise to two focal points on the normal geodesic to $h^{-1}(M)$ at (x, z) with corresponding principal curvatures

$$\lambda^+ = \cot(\theta/2), \quad \lambda^- = \cot((\theta + \pi)/2).$$

This completes the proof of the theorem. $\square$

We now construct the examples of Miyaoka and Ozawa. Recall that a compact proper Dupin hypersurface M in S^4 with two principal curvatures must be a cyclide of Dupin, that is, the image under a Möbius transformation of S^4 of a standard product of spheres,

$$S^1(r) \times S^2(s) \subset S^4(1) \subset \mathbf{R}^5, \quad r^2 + s^2 = 1.$$

A conformal, nonisometric image of an isoparametric cyclide does not have constant principal curvatures. Similarly, a compact proper Dupin hypersurface in S^4 with three principal curvatures must be Lie equivalent to an isoparametric hypersurface in S^4 with three principal curvatures, but it need not have constant principal curvatures itself.

Corollary 4.45. *Let M be a nonisoparametric compact, connected proper Dupin hypersurface embedded in S^4 with g principal curvatures, where $g = 2$ or 3. Then $h^{-1}(M)$ is a compact, connected proper Dupin hypersurface in S^7 with $2g$ principal curvatures that is not Lie equivalent to an isoparametric hypersurface in S^7.*

Proof. Suppose that $\lambda = \cot \theta$ and $\mu = \cot \alpha$ are two distinct nonconstant principal curvature functions on M. Let

$$\lambda^+ = \cot(\theta/2), \quad \lambda^- = \cot((\theta + \pi)/2),$$
$$\mu^+ = \cot(\alpha/2), \quad \mu^- = \cot((\alpha + \pi)/2),$$

be the four distinct principal curvature functions on $h^{-1}(M)$ induced from λ and μ. Then the Lie curvature

$$\Psi = \frac{(\lambda^+ - \lambda^-)(\mu^+ - \mu^-)}{(\lambda^+ - \mu^-)(\mu^+ - \lambda^-)} = \frac{2}{1 + \cos(\theta - \alpha)},$$

is not constant on $h^{-1}(M)$, and therefore $h^{-1}(M)$ is not Lie equivalent to an isoparametric hypersurface in S^7. $\square$

Certain parts of the construction of Miyaoka and Ozawa are also valid if **H** is replaced by the Cayley numbers or a more general Clifford algebra. See the paper of Miyaoka and Ozawa [121] for a discussion of this point.

As noted in Section 4.5, Miyaoka [113] proved that the assumption that the Lie curvature Ψ is constant on a compact, connected proper Dupin hypersurface M in S^n with four principal curvatures, together with an additional assumption regarding the intersections of leaves of the various principal foliations, implies that M is Lie equivalent to an isoparametric hypersurface. (Miyaoka [114] also proved a similar result for compact proper Dupin hypersurfaces with six principal curvatures.) Thorbergsson [190] (see also Stolz [177] and Grove–Halperin [79]) proved that for a compact proper Dupin hypersurface in S^n with four principal curvatures, the multiplicities of the principal curvatures must satisfy $m_1 = m_2, m_3 = m_4$, when the principal curvatures are appropriately ordered. Then Cecil, Chi and Jensen [41] used a different approach than Miyaoka to prove that if M is a compact, connected proper Dupin hypersurface in S^n with four principal curvatures whose multiplicities satisfy $m_1 = m_2 \geq 1, m_3 = m_4 = 1$ and constant Lie curvature Ψ, then M is Lie equivalent to an isoparametric hypersurface. Thus, Miyaoka's additional assumption regarding the intersections of leaves of the various principal foliations is not needed in that case. It remains an open question whether Miyaoka's additional assumption can be removed in the case where $m_3 = m_4$ is also allowed to be greater than one, although this has been conjectured to be true by Cecil and Jensen [45, pp. 3–4].

5

Dupin Submanifolds

In this chapter, we concentrate on local results that have been obtained using Lie sphere geometry. The main results presented here are the classification of proper Dupin submanifolds with two principal curvatures (cyclides of Dupin) in Section 5.4 and the classification of proper Dupin hypersurfaces with three principal curvatures in $\mathbf{R}^4$ in Section 5.7. To obtain these classifications, we develop the method of moving Lie frames which can be used in the further study of Dupin submanifolds, or more generally, Legendre submanifolds.

5.1 Local Constructions

Pinkall [150] introduced four constructions for obtaining a Dupin hypersurface W in $\mathbf{R}^{n+m}$ from a Dupin hypersurface M in $\mathbf{R}^n$. We first describe these constructions in the case $m = 1$ as follows.

Begin with a Dupin hypersurface M^{n-1} in $\mathbf{R}^n$ and then consider $\mathbf{R}^n$ as the linear subspace $\mathbf{R}^n \times \{0\}$ in $\mathbf{R}^{n+1}$. The following constructions yield a Dupin hypersurface W^n in $\mathbf{R}^{n+1}$.

(1) Let W^n be the cylinder $M^{n-1} \times \mathbf{R}$ in $\mathbf{R}^{n+1}$.
(2) Let W^n be the hypersurface in $\mathbf{R}^{n+1}$ obtained by rotating M^{n-1} around an axis $\mathbf{R}^{n-1} \subset \mathbf{R}^n$.
(3) Let W^n be a tube in $\mathbf{R}^{n+1}$ around M^{n-1}.
(4) Project M^{n-1} stereographically onto a hypersurface $V^{n-1} \subset S^n \subset \mathbf{R}^{n+1}$. Let W^n be the cone over V^{n-1} in $\mathbf{R}^{n+1}$.

In general, these constructions introduce a new principal curvature of multiplicity one which is constant along its lines of curvature. The other principal curvatures are determined by the principal curvatures of M^{n-1}, and the Dupin property is preserved for these principal curvatures. These constructions can be generalized to produce a new principal curvature of multiplicity m by considering $\mathbf{R}^n$ as a subset of $\mathbf{R}^n \times \mathbf{R}^m$ rather than $\mathbf{R}^n \times \mathbf{R}$.

Although Pinkall gave these four constructions, his [150, Theorem 4, p. 438] showed that the cone construction is redundant, since it is Lie equivalent to a tube. This will be explained further in Remark 5.13 (p. 144). For this reason, we will only study three standard constructions: tubes, cylinders and surfaces of revolution in the next section.

A Dupin submanifold obtained from a lower-dimensional Dupin submanifold via one of these standard constructions is said to be *reducible*. More generally, a Dupin submanifold which is locally Lie equivalent to such a Dupin submanifold is called *reducible*.

Using these constructions, Pinkall was able to produce a proper Dupin hypersurface in Euclidean space with an arbitrary number of distinct principal curvatures, each with any given multiplicity (see Theorem 5.1 below). In general, these proper Dupin hypersurfaces cannot be extended to compact Dupin hypersurfaces without losing the property that the number of distinct principal curvatures is constant (see Section 5.2). For now, we give a proof of Pinkall's theorem without attempting to compactify the hypersurfaces constructed.

Theorem 5.1. *Given positive integers $v_1, \ldots, v_g$ with*

$$v_1 + \cdots + v_g = n - 1,$$

there exists a proper Dupin hypersurface in $\mathbf{R}^n$ with g distinct principal curvatures having respective multiplicities $v_1, \ldots, v_g$.

Proof. The proof is by an inductive construction, which will be clear once the first few examples are done. To begin, note that a usual torus of revolution in $\mathbf{R}^3$ is a proper Dupin hypersurface with two principal curvatures. To construct a proper Dupin hypersurface M^3 in $\mathbf{R}^4$ with three principal curvatures, each of multiplicity one, begin with an open subset U of a torus of revolution in $\mathbf{R}^3$ on which neither principal curvature vanishes. Take M^3 to be the cylinder $U \times \mathbf{R}$ in $\mathbf{R}^3 \times \mathbf{R} = \mathbf{R}^4$. Then M^3 has three distinct principal curvatures at each point, one of which is zero. These are clearly constant along their corresponding one-dimensional curvature surfaces (lines of curvature).

To get a proper Dupin hypersurface in $\mathbf{R}^5$ with three principal curvatures having respective multiplicities $v_1 = v_2 = 1$, $v_3 = 2$, one simply takes

$$U \times \mathbf{R}^2 \subset \mathbf{R}^3 \times \mathbf{R}^2 = \mathbf{R}^5.$$

To obtain a proper Dupin hypersurface M^4 in $\mathbf{R}^5$ with four principal curvatures, first invert the hypersurface M^3 above in a 3-sphere in $\mathbf{R}^4$, chosen so that the image of M^3 contains an open subset W^3 on which no principal curvature vanishes. The hypersurface W^3 is proper Dupin, since the proper Dupin property is preserved by Möbius transformations. Now take M^4 to be the cylinder $W \times \mathbf{R}$ in $\mathbf{R}^4 \times \mathbf{R} = \mathbf{R}^5$. $\square$

We now turn to a full discussion of Pinkall's constructions in the setting of Lie sphere geometry.

5.2 Reducible Dupin Submanifolds

In this section, we study the standard constructions, introduced by Pinkall [150], for obtaining a proper Dupin submanifold μ with $g + 1$ distinct curvature spheres from a lower-dimensional proper Dupin submanifold λ with g curvature spheres in detail. Pinkall only described these constructions locally, i.e., he began with a hypersurface M^{n-1} embedded in $\mathbf{R}^n$. Here we formulate the constructions in the context of Lie sphere geometry. In each construction, we imitate the case where the Euclidean projection of λ is an immersion, but we do not assume this. We then determine the curvature spheres of μ and their multiplicities. Although this approach is more complicated than simply working in $\mathbf{R}^n$, it enables us to answer important questions concerning the possibility of constructing compact proper Dupin submanifolds by these methods.

We first set some notation common to all three constructions. Let

$$\{e_1, \ldots, e_{n+m+3}\}$$

be the standard orthonormal basis for $\mathbf{R}_2^{n+m+3}$, with e_1 and e_{n+m+3} timelike. Let $\mathbf{P}^{n+m+2}$ be the projective space determined by $\mathbf{R}_2^{n+m+3}$, with corresponding Lie quadric Q^{n+m+1}. Let

$$\mathbf{R}_2^{n+3} = \mathrm{Span}\{e_1, \ldots, e_{n+2}, e_{n+m+3}\},$$

and let $\mathbf{P}^{n+2}$ and Q^{n+1} be the corresponding projective space and Lie quadric, respectively. Let Λ^{2n-1} and $\Lambda^{2(n+m)-1}$ be the spaces of projective lines on Q^{n+1} and Q^{n+m+1}, respectively. Finally, let $u_k = e_{k+2}$, $1 \leq k \leq n + m$, and

$$\mathbf{R}^n = \mathrm{Span}\{e_3, \ldots, e_{n+2}\} = \mathrm{Span}\{u_1, \ldots, u_n\}, \tag{5.1}$$
$$\mathbf{R}^{n+m} = \mathrm{Span}\{e_3, \ldots, e_{n+m+2}\} = \mathrm{Span}\{u_1, \ldots, u_{n+m}\}.$$

A. Tubes

We will construct a Legendre submanifold which corresponds to building a tube in $\mathbf{R}^{n+m}$ around an $(n - 1)$-dimensional submanifold in $\mathbf{R}^n$. This can be done for any Legendre submanifold λ, although we will assume that λ is a proper Dupin submanifold. We will work with Euclidean projections of the Legendre submanifolds here, but one could just as well use spherical projections and construct a tube using the spherical metric (see Remark 5.13, p. 144).

Let $\lambda : M^{n-1} \to \Lambda^{2n-1}$ be a proper Dupin submanifold with g distinct curvature spheres whose locus of point spheres does not contain the improper point $[e_1 - e_2]$. Then the point sphere map $[k_1]$ and the hyperplane map $[k_2]$ for λ can be written as follows:

$$k_1 = (1 + f \cdot f, 1 - f \cdot f, 2f, 0)/2, \qquad k_2 = (f \cdot \xi, -f \cdot \xi, \xi, 1). \tag{5.2}$$

These equations define the Euclidean projection f and Euclidean field of unit normals ξ for λ. As the calculations to follow will show, in order to construct a proper Dupin submanifold, we need to assume that f has constant rank. We will distinguish the case where f is an immersion from the case where f has lower rank.

First, we assume that f is an immersion. Thus, we can think of

$$f : M^{n-1} \rightarrow \mathbf{R}^n$$

as an oriented hypersurface with field of unit normals ξ. The domain of the Legendre submanifold μ, corresponding to a tube of radius ε over f, is the unit normal bundle B^{n+m-1} to $f(M^{n-1})$ in $\mathbf{R}^{n+m}$. The normal vector fields,

$$\xi, u_{n+1}, \dots, u_{n+m},$$

are all parallel with respect to the normal connection of $f(M^{n-1})$ in $\mathbf{R}^{n+m}$. This enables us to define a global trivialization of B^{n+m-1} with the properties of the local trivialization used in Section 4.3. Specifically, let

$$S^m = \{(y_0, \dots, y_m) \mid y_0^2 + \cdots + y_m^2 = 1\}. \tag{5.3}$$

Then the map $(x, y) \mapsto (x, \eta(x, y))$ with

$$\eta(x, y) = y_0 \xi(x) + y_1 u_{n+1} + \cdots + y_m u_{n+m}, \tag{5.4}$$

is a diffeomorphism from $M^{n-1} \times S^m \rightarrow B^{n+m-1}$.

We now define the map

$$\mu : M^{n-1} \times S^m \rightarrow \Lambda^{2(n+m)-1}, \tag{5.5}$$

corresponding to the tube of radius ε around $f(M^{n-1})$ in $\mathbf{R}^{n+m}$. This construction works for any real number ε. In particular, the case $\varepsilon = 0$ is just the Legendre submanifold induced by the immersion $f(M^{n-1})$ as a submanifold of codimension $m + 1$ in $\mathbf{R}^{n+m}$.

The map μ is defined by its Euclidean projection F and its Euclidean field of unit normals η as in equation (5.4), both of which are maps from $M^{n-1} \times S^m$ into $\mathbf{R}^{n+m}$. For $x \in M^{n-1}$ and $y = (y_0, \dots, y_m) \in S^m$, we define the map μ by the formula

$$\mu(x, y) = [K_1(x, y), K_2(x, y)],$$

where

$$K_1 = (1 + F \cdot F, 1 - F \cdot F, 2F, 0)/2, \qquad K_2 = (F \cdot \eta, -F \cdot \eta, \eta, 1), \tag{5.6}$$

with

$$F(x, y) = f(x) + \varepsilon(y_0 \xi(x) + y_1 u_{n+1} + \cdots + y_m u_{n+m}), \tag{5.7}$$

and η is given by equation (5.4). The image of the map $F : M^{n-1} \times S^m \rightarrow \mathbf{R}^{n+m}$ is the tube of radius ε over the submanifold $f(M^{n-1})$ in $\mathbf{R}^{n+m}$, and η is a field of unit normals to this tube. The map $[K_1]$ is the point sphere map for μ, and $[K_2]$ is the hyperplane map for μ.

To see that μ is a Legendre submanifold, we must check the Legendre conditions (1)–(3) of Theorem 4.3 in Chapter 4, p. 59. The Legendre condition (1) is easily verified. To check conditions (2) and (3) and find the curvature spheres, we must compute the differentials of K_1 and K_2. We decompose the tangent space to $M^{n-1} \times S^m$ at the point $p = (x, y)$ as

$$T_p(M^{n-1} \times S^m) = T_x M^{n-1} \times T_y S^m, \tag{5.8}$$

and denote a typical tangent vector by (X, Y). For K_1 and K_2 as in equation (5.6), it is easy to check that for real numbers r and s, at least one of which is not zero,

$$d(rK_1 + sK_2)(X, Y) \in [K_1, K_2] \Leftrightarrow d(rF + s\eta)(X, Y) = 0. \tag{5.9}$$

The tangent space in equation (5.8) has a basis of vectors of the form $(X, 0)$ and $(0, Y)$, where

$$Y = (Y_0, \dots, Y_m) \in T_y S^m.$$

From equations (5.4) and (5.7), we compute

$$dF(X, 0) = df(X) + \varepsilon y_0 \, d\xi(X), \tag{5.10}$$
$$d\eta(X, 0) = y_0 \, d\xi(X), \tag{5.11}$$
$$dF(0, Y) = \varepsilon(Y_0 \xi(x) + Y_1 u_{n+1} + \cdots + Y_m u_{n+m}), \tag{5.12}$$
$$d\eta(0, Y) = Y_0 \xi(x) + Y_1 u_{n+1} + \cdots + Y_m u_{n+m}. \tag{5.13}$$

The Legendre contact condition (3) reduces to $dF \cdot \eta = 0$ for K_1 and K_2 as in equation (5.6). This can now be checked directly from equations (5.4), (5.10) and (5.12), using the fact that $df(X)$ and $d\xi(X)$ are both orthogonal to $\xi(x)$ and that

$$y_0 Y_0 + \cdots + y_m Y_m = 0.$$

Next we locate the curvature spheres of μ. The fact that the Legendre condition (2) is satisfied follows from these calculations. From equations (5.12) and (5.13), we see that

$$d(F - \varepsilon \eta)(0, Y) = 0,$$

for every $Y \in T_y S^m$. Thus, $[K_1 - \varepsilon K_2]$ is a curvature sphere of multiplicity at least m at each point of $M^{n-1} \times S^m$. From formulas (2.14) and (2.16) (p. 16) of Chapter 2, we see that $[K_1 - \varepsilon K_2]$ represents an oriented hypersphere with center $f(x)$ and radius $-\varepsilon$ (the minus sign is due to the outward normal). This is the new family of curvature spheres which results from this construction. Note that if there were a nonzero vector $X \in T_x M^{n-1}$ such that $df(X) = 0$, then we would have

$$d(F - \varepsilon \eta)(X, 0) = 0$$

also, and the curvature sphere $[K_1 - \varepsilon K_2]$ would have multiplicity $m + v$ at (x, y), where v is the nullity of df at x. This shows that f must have constant rank for this construction to yield a proper Dupin submanifold. Since we have assumed that f

is an immersion, the curvature sphere $[K_1 - \varepsilon K_2]$ has constant multiplicity m. The curvature surfaces of $[K_1 - \varepsilon K_2]$ are of the form $\{x\} \times S^m$, for $x \in M^{n-1}$. From equation (2.14) of Chapter 2, p. 16, the analytic condition for these curvature spheres to have radius $-\varepsilon$ is

$$\langle K_1 - \varepsilon K_2, \varepsilon e_1 - \varepsilon e_2 + e_{n+m+3} \rangle = 0. \tag{5.14}$$

The centers of these spheres all lie in $\mathbf{R}^n$, i.e.,

$$\langle K_1 - \varepsilon K_2, u_{n+i} \rangle = \langle K_1 - \varepsilon K_2, e_{n+2+i} \rangle = 0, \quad 1 \le i \le m.$$

Thus, the image of the map $[K_1 - \varepsilon K_2]$ is contained in the $(n+1)$-dimensional linear subspace $E \subset \mathbf{P}^{n+m+2}$ whose orthogonal complement is

$$E^\perp = \mathrm{Span}\{e_{n+3}, \dots, e_{n+m+2}, \varepsilon e_1 - \varepsilon e_2 + e_{n+m+3}\}, \tag{5.15}$$

which has signature $(m, 1)$.

The computation to determine the other curvature spheres splits into two cases depending on whether or not the coordinate y_0 of y is zero:

CASE 1. $y_0 \ne 0$.

This is the case where the vector η in equation (5.4) is not orthogonal to $\mathbf{R}^n$. Then from equations (5.10)–(5.11), we have for any real numbers r and s,

$$d(rF + s\eta)(X, 0) = r\, df(X) + y_0(r\varepsilon + s)\, d\xi(X).$$

Since $y_0 \ne 0$, we can obtain all possible linear combinations of $df(X)$ and $d\xi(X)$ by appropriate choices of r and s. Thus, there exist real numbers r and s such that

$$d(rF + s\eta)(X, 0) = 0,$$

precisely when $df(X)$ and $d\xi(X)$ are linearly dependent, i.e., X is a principal vector of the original Legendre submanifold λ. Hence the other curvature spheres, with their respective multiplicities, correspond to the curvature spheres of λ at x. Since λ is Dupin, these curvature spheres of μ are constant along their curvature surfaces, which have the form $S \times \{y\}$, where S is a curvature surface of λ through x. The number $\gamma(x, y)$ of curvature spheres of μ at the point (x, y) equals $g + 1$, where g is the number of curvature spheres of λ.

CASE 2. $y_0 = 0$.

This is the case when η is orthogonal to $\mathbf{R}^n$. At these points, we have from equations (5.10)–(5.11) that

$$dF(X, 0) = df(X), \qquad d\eta(X, 0) = 0. \tag{5.16}$$

Formula (5.16) implies that $[K_2]$ is a curvature sphere of multiplicity $n - 1$ at the point (x, y). Hence, at (x, y) there are only two distinct curvature spheres, $[K_2]$ and $[K_1 - \varepsilon K_2]$, having respective multiplicities $n - 1$ and m. These curvature spheres are constant along their curvature surfaces, which have the form $M^{n-1} \times \{y\}$ and $\{x\} \times S^m$, respectively. This holds regardless of the original Legendre submanifold λ. In terms of Euclidean geometry, the points satisfying $y_0 = 0$ correspond to points at a distance ε away from the original space $\mathbf{R}^n$. See Example 4.10 of Chapter 4, p. 69, of a tube over a torus

$$T^2 \subset \mathbf{R}^3 \subset \mathbf{R}^4.$$

The set of points where $y_0 = 0$ is diffeomorphic to $M^{n-1} \times S^{m-1}$. We summarize our results for the tube construction in the following proposition.

Proposition 5.2. *Suppose that* $\lambda : M^{n-1} \to \Lambda^{2n-1}$ *is a proper Dupin submanifold with g distinct curvature spheres such that the Euclidean projection f is an immersion of* M^{n-1} *into* $\mathbf{R}^n \subset \mathbf{R}^{n+m}$. *Then the tube construction yields a Dupin submanifold* μ *defined on the unit normal bundle* B^{n+m-1} *of* $f(M^{n-1})$ *in* $\mathbf{R}^{n+m}$. *The number* $\gamma(x, \eta)$ *of distinct curvature spheres of* μ *at a point* $(x, \eta) \in B^{n+m-1}$ *is as follows:*

(a) $\gamma(x, \eta) = 2$, *if* η *is orthogonal to* $\mathbf{R}^n$ *in* $\mathbf{R}^{n+m}$.
(b) $\gamma(x, \eta) = g + 1$ *otherwise.*

Remark 5.3. In the case $\varepsilon = 0$, μ is the Legendre submanifold induced by the immersion $f(M^{n-1})$ as a submanifold of codimension $m + 1$ in $\mathbf{R}^{n+m}$. Theorem 4.15 of Chapter 4, p. 74, describes the curvature spheres of μ. The point sphere map $[K_1]$ in equation (5.6) is a curvature sphere of multiplicity m, which lies in the $(n + 1)$-dimensional subspace E with orthogonal complement $E^\perp$ in equation (5.15) with $\varepsilon = 0$. The tubes of radius $\varepsilon \neq 0$ over $f(M^{n-1})$ are parallel submanifolds of μ.

We assume now that the Euclidean projection f of λ has constant rank less that $n - 1$. Then λ is the Legendre submanifold induced by an immersed submanifold $\phi : V \to \mathbf{R}^n$ of codimension $\nu + 1$, and the domain of λ is the unit normal bundle B^{n-1} of $\phi(V)$ in $\mathbf{R}^n$. As in Remark 5.3, we first consider the Legendre submanifold μ induced by the submanifold $\phi(V)$ of codimension $m + \nu + 1$ in $\mathbf{R}^{n+m}$. The number of distinct curvature spheres of μ at a point (x, η) in the unit normal bundle B^{n+m+1} is determined by Theorem 4.15 of Chapter 4, p. 74. We decompose the vector η as follows:

$$\eta = \cos\theta\, \xi + \sin\theta\, u, \quad 0 \leq \theta \leq \pi/2,$$

where ξ is a unit vector in $\mathbf{R}^n$ normal to $\phi(V)$ at $\phi(x)$, and u is a unit vector orthogonal to $\mathbf{R}^n$ in $\mathbf{R}^{n+m}$. Since the shape operator $A_u = 0$, we have $A_\eta = \cos\theta\, A_\xi$. If $\cos\theta$ is nonzero, then the number of distinct curvature spheres of μ at (x, η) is the same as the number of distinct curvature spheres of λ at (x, ξ), since the point sphere map is a curvature sphere in both cases. On the other hand, if $\cos\theta = 0$, then we have $A_\eta = A_u = 0$, and the number of distinct curvature spheres of μ at (x, η) is two. This is similar to case (a) in Proposition 5.2.

Using the local trivialization of B^{n+m-1} given in Section 4.3, it is easy to check that μ is Dupin, since we are assuming that λ is Dupin. Furthermore, the point sphere map $[K_1]$ is a curvature sphere of μ of multiplicity $m + \nu$, and it lies in an $(n + 1)$-dimensional subspace E of $\mathbf{P}^{n+m+2}$ whose orthogonal complement is the space $E^\perp$ in equation (5.15) with $\varepsilon = 0$.

The Legendre submanifold corresponding to a tube of radius $\varepsilon \neq 0$ over $\phi(V)$ in $\mathbf{R}^{n+m}$ is a parallel submanifold to μ. Thus, it is also Dupin, and it has the same number of curvature spheres at each point as μ. We summarize these results in the following proposition.

Proposition 5.4. *Suppose that* $\lambda : B^{n-1} \to \Lambda^{2n-1}$ *is a proper Dupin submanifold with g distinct curvature spheres induced by an immersed submanifold* $\phi(V)$ *of codimension* $\nu + 1$ *in* $\mathbf{R}^n \subset \mathbf{R}^{n+m}$. *Then the tube construction yields a Dupin submanifold* μ *defined on the unit normal bundle* B^{n+m-1} *to* $\phi(V)$ *in* $\mathbf{R}^{n+m}$. *The number* $\gamma(x, \eta)$ *of distinct curvature spheres of* μ *at a point* $(x, \eta) \in B^{n+m-1}$ *is as follows:*

(a) $\gamma(x, \eta) = 2$, *if* η *is orthogonal to* $\mathbf{R}^n$ *in* $\mathbf{R}^{n+m}$.
(b) $\gamma(x, \eta) = g$ *otherwise.*

Remark 5.5. The original purpose of Pinkall's constructions was to increase the number of distinct curvature spheres by one, as in Proposition 5.2. However, as Proposition 5.4 shows, this does not happen when λ is the Legendre submanifold induced by a submanifold $\phi(V)$ of codimension greater than in one in $\mathbf{R}^n$. Still we consider the Dupin submanifold μ in Proposition 5.4 to be reducible, since it is obtained from λ by one of the standard constructions. The following is a concrete example of this phenomenon.

Example 5.6 (tube over a Veronese surface in $S^4 \subset S^5$). We consider the case where V^2 is a Veronese surface embedded in $S^4 \subset S^5$, where S^4 is a great sphere in S^5. We first recall the details of the Veronese surface. Let S^2 be the unit sphere in $\mathbf{R}^3$ given by the equation
$$u^2 + v^2 + w^2 = 1.$$
Consider the map from S^2 into the unit sphere $S^4 \subset \mathbf{R}^5$ given by

$$(u, v, w) \mapsto \left(\sqrt{3}vw, \sqrt{3}wu, \sqrt{3}uv, \frac{\sqrt{3}}{2}(u^2 - v^2), w^2 - \frac{u^2 + v^2}{2} \right).$$

This map takes the same value on antipodal points of S^2, so it induces a map $\phi : \mathbf{P}^2 \to S^4$, and one can show that ϕ is an embedding. The surface $V^2 = \phi(\mathbf{P}^2)$ is called a *Veronese surface.* One can show (see, for example, [52, Example 7.3, pp. 296–299]) that a tube over V^2 of radius ε, for $0 < \varepsilon < \pi/3$, in the spherical metric of S^4 is an isoparametric hypersurface M^3 with $g = 3$ distinct principal curvatures. This isoparametric hypersurface M^3 is not reducible, because the Veronese surface is substantial (does not lie in a hyperplane) in $\mathbf{R}^5$, so M^3 is not obtained as a result of the tube construction as described above. (See Takeuchi [181] for further discussion of proper Dupin hypersurfaces obtained as tubes over symmetric submanifolds of codimension greater than one in space-forms.)

Now embed $\mathbf{R}^5$ as a hyperplane through the origin in $\mathbf{R}^6$ and let e_6 be a unit normal vector to $\mathbf{R}^5$ in $\mathbf{R}^6$. The surface V^2 is a subset of the unit sphere $S^5 \subset \mathbf{R}^6$. As in the calculations made prior to Proposition 5.4, one can show that a tube over V^2 of radius ε in S^5 is not an isoparametric hypersurface, nor is it even a proper Dupin hypersurface, because the number of distinct principal curvatures is not constant on the unit normal bundle B^4 to V^2 in S^5. Specifically, if μ is the Legendre submanifold induced by the submanifold $V^2 \subset S^5$, then μ has two distinct curvature spheres at points in B^4 of the form $(x, \pm e_6)$, and three distinct curvature spheres at all other points of B^4. A tube W^4 over V^2 in S^5 is a reducible Dupin hypersurface, but it is not proper Dupin. At points of W^4 corresponding to the points $(x, \pm e_6)$ in B^4, there are two principal curvatures, both of multiplicity two. At the other points of W^4, there are three distinct principal curvatures, one of multiplicity two, and the others of multiplicity one. Thus, W^4 has an open dense subset U which is a reducible proper Dupin hypersurface with three principal curvatures at each point, but W^4 itself is not proper Dupin.

B. Cylinders

As before, we begin with a proper Dupin submanifold $\lambda : M^{n-1} \to \Lambda^{2n-1}$ with g distinct curvature spheres at each point, and assume that the locus of point spheres does not contain the improper point $[e_1 - e_2]$. We can write the point sphere map $[k_1]$ and the hyperplane map $[k_2]$ in the form of equation (5.2), and thereby define the Euclidean projection f and the Euclidean field of unit normals ξ as maps from M^{n-1} to $\mathbf{R}^n$. Usually, one thinks of the cylinder built over f in $\mathbf{R}^{n+m} = \mathbf{R}^n \times \mathbf{R}^m$ to be the map from $M^{n-1} \times \mathbf{R}^m$ to $\mathbf{R}^{n+m}$ given by

$$(x, z) \mapsto f(x) + z.$$

Here we attempt to extend this map to a map defined on $M^{n-1} \times S^m$ by working in the context of Lie sphere geometry. This is accomplished by mapping all points in the set $M^{n-1} \times \{\infty\}$ to the improper point in Lie sphere geometry. The Legendre immersion condition (2) of Theorem 4.3 in Chapter 4, p. 59, can still be satisfied at points of the form (x, ∞) because the normal vector varies as x varies. However, as the computations below show, the Legendre immersion condition (2) is only satisfied at points of the form (x, ∞) for which the map ξ has rank $n - 1$ at x.

As in the tube construction, we consider S^m as in equation (5.3). We relate S^m to $\mathbf{R}^m$ via stereographic projection

$$\tau : S^m - \{P\} \to \mathbf{R}^m, \text{ where } P = (-1, 0, \ldots, 0),$$

given by

$$\tau(y_0, \ldots, y_m) = \frac{1}{1 + y_0}(y_1, \ldots, y_m) = (z_1, \ldots, z_m) = z.$$

We define the Legendre submanifold μ corresponding to the cylinder over f in $\mathbf{R}^{n+m}$ by giving its Euclidean projection F and Euclidean field of unit normals η. For a cylinder defined on $M^{n-1} \times \mathbf{R}^m$, these should obviously be defined as follows:

$$F(x, z) = f(x) + z_1 u_{n+1} + \cdots + z_m u_{n+m}, \tag{5.17}$$

$$\eta(x, z) = \xi(x). \tag{5.18}$$

We can now obtain the extension to $M^{n-1} \times \{P\}$ by writing the maps K_1 and K_2 induced from F and η in the usual way given in equation (5.6). First note that

$$F \cdot F = f \cdot f + \sum_{i=1}^{m} z_i^2 = f \cdot f + \frac{1 - y_0^2}{(1 + y_0)^2} = f \cdot f + \frac{1 - y_0}{1 + y_0}.$$

We can multiply K_1 by $2(1 + y_0)$ and get that $[K_1]$ equals

$$[(2 + f \cdot f(1 + y_0), 2y_0 - f \cdot f(1 + y_0), 2(1 + y_0)f + 2y_1 u_{n+1} + \cdots + 2y_m u_{n+m}, 0)].$$

When the values $y_0 = -1$ and $y_i = 0$, $1 \le i \le m$, are substituted into this formula, we get

$$[K_1] = [(1, -1, 0, \ldots, 0)],$$

the improper point, as desired.

Since the formula for $K_2(x, y)$ does not depend on the point $y \in S^m$, it does not need to be modified to incorporate the special point P. Specifically,

$$K_2(x, y) = (f(x) \cdot \xi(x), -f(x) \cdot \xi(x), \xi(x), 1).$$

As with the tube construction, it is easy to check that K_1 and K_2 satisfy the Legendre condition (1) of Theorem 4.3 in Chapter 4, p. 59. To check the Legendre conditions (2) and (3) and to locate the curvature spheres, we compute the differentials of K_1 and K_2. We first consider points (x, y) with $y_0 \ne -1$, that is $y \ne P$. For these points, it is simpler to use formulas (5.17)–(5.18) defined on $M^{n-1} \times \mathbf{R}^m$. Thus, we identify a point $y \in S^m - \{P\}$ with the point $z = \tau(y)$ in $\mathbf{R}^m$, and identify $T_y S^m$ with $T_z \mathbf{R}^m$ via $d\tau$. We write a typical tangent vector as (X, Z), for $X \in T_x M^{n-1}$ and $Z \in T_z \mathbf{R}^m$. As before, the location of the curvature spheres is determined by the condition

$$d(rF + s\eta)(X, Z) = 0.$$

If we write $Z = (Z_1, \ldots, Z_m)$, then we can compute from equations (5.17)–(5.18) that

$$dF(X, 0) = df(X), \qquad\qquad d\eta(X, 0) = d\xi(X), \tag{5.19}$$
$$dF(0, Z) = Z_1 u_{n+1} + \cdots + Z_m u_{n+m}, \qquad d\eta(0, Z) = 0. \tag{5.20}$$

The Legendre condition (3),

$$dF \cdot \eta = 0,$$

is easily verified using equations (5.18)–(5.20) and the Legendre condition (3) for λ, that is,

$$df \cdot \xi = 0.$$

From equation (5.20), it follows immediately that $[K_2]$ is a curvature sphere of multiplicity at least m at each point of $M^{n-1} \times \mathbf{R}^m$. This is also true on $M^{n-1} \times S^m$, since $[K_2]$ does not depend on the point $y \in S^m$. At each point, $[K_2]$ corresponds to a hyperplane in $\mathbf{R}^{n+m}$ in oriented contact with the cylinder along one of its rulings. Note further that if there is a nonzero vector $X \in T_x M^{n-1}$ such that $d\xi(X) = 0$, then $d\eta(X, 0) = 0$ also, and the curvature sphere $[K_2]$ has multiplicity $m + \nu$, where ν is the nullity of $d\xi$ at x. Thus, $[K_2]$ does not have constant multiplicity unless $d\xi$ has constant rank on M^{n-1}.

Therefore, we now assume that ξ has constant rank on M^{n-1}. We first consider the case when the rank of ξ is $n - 1$, i.e., ξ is an immersion on M^{n-1}. Then $[K_2]$ has multiplicity m and is constant along its curvature surfaces, which have the form $\{x\} \times S^m$. Since $[K_2]$ represents a hyperplane, it satisfies the equation

$$\langle K_2, e_1 - e_2 \rangle = 0.$$

Furthermore, the normal $\eta(x, y) = \xi(x)$ to the hyperplane lies in $\mathbf{R}^n$, and so

$$\langle K_2, u_{n+i} \rangle = \langle K_2, e_{n+2+i} \rangle = 0, \quad 1 \le i \le m.$$

Hence, the image of the map $[K_2]$ is contained in the $(n + 1)$-dimensional linear subspace E in $\mathbf{P}^{n+2}$ whose orthogonal complement is the $(m+1)$-dimensional vector space,

$$E^\perp = \mathrm{Span}\{e_{n+3}, \ldots, e_{n+m+2}, e_1 - e_2\}, \tag{5.21}$$

on which the scalar product $\langle , \rangle$ has signature $(m, 0)$.

From equation (5.19), we see that the other curvature spheres correspond exactly to the curvature spheres of the original Dupin submanifold λ. These curvature spheres are also constant along their curvature surfaces, which are of the form $S \times \{y\}$, where S is a curvature surface of λ. Thus, μ has $g + 1$ curvature spheres at the points (x, y) with $y \ne P$.

We now consider points in $M^{n-1} \times \{P\}$. Let x be an arbitrary point of M^{n-1}. We know that

$$dK_1(X, 0) = 0$$

at (x, P) for all $X \in T_x M^{n-1}$, since $[K_1]$ is constant on the set $M^{n-1} \times \{P\}$. On the other hand, one can easily compute that $dK_1(0, Y)$ is not in $[K_1, K_2]$ for any Y. Thus, $[K_1]$ is a curvature sphere of multiplicity $n - 1$ at each point of the form (x, P). By differentiating the formula for K_2 and using the fact that $df \cdot \xi = 0$, we get

$$dK_2(X, 0) = (f \cdot d\xi(X), -f \cdot d\xi(X), d\xi(X), 0). \tag{5.22}$$

Since we have assumed that ξ is an immersion, $dK_2(X, 0) \ne 0$ for $X \ne 0$. This shows that the Legendre condition (2) is satisfied and that there are exactly two distinct curvature spheres, $[K_1]$ with multiplicity $n - 1$ and $[K_2]$ with multiplicity m at (x, P). These have respective curvature surfaces $M^{n-1} \times \{P\}$ and $\{x\} \times S^m$. The Dupin condition is clearly satisfied by these curvature surfaces, and thus μ is Dupin. We summarize these results as follows.

Proposition 5.7. *Suppose that $\lambda : M^{n-1} \to \Lambda^{2n-1}$ is a proper Dupin submanifold with g distinct curvature spheres such that the Euclidean field of unit normals ξ is* *an immersion. Then the cylinder construction yields a Dupin submanifold μ defined on $M^{n-1} \times S^m$. The number $\gamma(x, y)$ of distinct curvature spheres of μ at a point $(x, y) \in M^{n-1} \times S^m$ is as follows:*

(a) $\gamma(x, y) = 2$, if y is the pole P of the stereographic projection τ from S^m to $\mathbf{R}^m$.
(b) $\gamma(x, y) = g + 1$ otherwise.

The cylinder construction also yields a Dupin submanifold defined on

$$M^{n-1} \times \mathbf{R}^m,$$

if ξ has constant rank $n - 1 - \nu$, for $\nu \geq 1$. However, the construction does not extend to $M^{n-1} \times \{P\}$ because the Legendre immersion condition (2) is not satisfied at those points. Specifically, if $X \in T_x M^{n-1}$ is a nonzero vector such that $d\xi(X) = 0$, then $dK_1(X, 0)$ and $dK_2(X, 0)$ are both zero at the point (x, P). Furthermore, the number of distinct curvature spheres of the cylinder μ on $M^{n-1} \times \mathbf{R}^m$ is g, not $g + 1$, since the hyperplane map is already a curvature sphere of λ. The curvature surfaces of $[K_2]$ are of the form $S \times S^m$, where S is a curvature surface of λ. Thus, we have the following.

Proposition 5.8. *Suppose that $\lambda : M^{n-1} \to \Lambda^{2n-1}$ is a proper Dupin submanifold with g distinct curvature spheres such that the Euclidean field of unit normals ξ has constant rank $n - 1 - \nu$, where $\nu \geq 1$. Then the cylinder construction yields a Dupin submanifold μ defined on $M^{n-1} \times \mathbf{R}^m$ with g distinct curvature spheres at each point.*

C. Surfaces of Revolution

As before, we begin with a proper Dupin submanifold $\lambda : M^{n-1} \to \Lambda^{2n-1}$ with g distinct curvature spheres at each point, and assume that the point sphere map $[k_1]$ does not contain the improper point. We write the point sphere map $[k_1]$ and the hyperplane map $[k_2]$ in the form of equation (5.2), and thereby define the Euclidean projection f and the Euclidean field of unit normals ξ as maps from M^{n-1} to $\mathbf{R}^n$. We now want to construct the Legendre submanifold μ obtained by revolving the "profile submanifold" f around an "axis of revolution"

$$\mathbf{R}^{n-1} \subset \mathbf{R}^n \subset \mathbf{R}^{n+m}.$$

We do not assume that f is an immersion, nor that the image of f is disjoint from the axis $\mathbf{R}^{n-1}$. For simplicity, we assume that the axis goes through the origin in $\mathbf{R}^n$ and that the standard orthonormal vectors $u_1, \ldots, u_{n+m}$ have been chosen so that

$$\mathbf{R}^{n-1} = \mathrm{Span}\{u_1, \ldots, u_{n-1}\} = \mathrm{Span}\{e_3, \ldots, e_{n+1}\}.$$

We write the sphere S^m in the form

$$S^m = \{y = y_0 u_n + y_1 u_{n+1} + \cdots + y_m u_{n+m} \mid y_0^2 + \cdots + y_m^2 = 1\}.$$

We can now define the new Legendre submanifold,

$$\mu : M^{n-1} \times S^m \to \Lambda^{2(n+m)-1},$$

by giving its Euclidean projection F and Euclidean field of unit normals η. First, we decompose the maps f and ξ into components along $\mathbf{R}^{n-1}$ and orthogonal to $\mathbf{R}^{n-1}$ in $\mathbf{R}^n$ and write

$$f(x) = \hat{f}(x) + f_n(x)u_n, \quad \hat{f}(x) \in \mathbf{R}^{n-1}, \tag{5.23}$$

$$\xi(x) = \hat{\xi}(x) + \xi_n(x)u_n, \quad \hat{\xi}(x) \in \mathbf{R}^{n-1}. \tag{5.24}$$

For $x \in M^{n-1}$, $y \in S^m$, we let

$$F(x, y) = \hat{f}(x) + f_n(x)y, \tag{5.25}$$

$$\eta(x, y) = \hat{\xi}(x) + \xi_n(x)y. \tag{5.26}$$

Note that for $y = u_n$, we have

$$F(x, u_n) = f(x), \qquad \eta(x, u_n) = \xi(x),$$

that is, we have the profile submanifold for the surface of revolution.

For a fixed point $x \in M^{n-1}$, the points $F(x, y)$, $y \in S^m$, form an m-dimensional sphere of radius $|f_n(x)|$ obtained by revolving the point $f(x)$ about the axis $\mathbf{R}^{n-1}$ in $\mathbf{R}^{n+m}$, provided that $f_n(x)$ is not zero.

The maps K_1 and K_2 are then defined by equation (5.6) as before. Again, it is easy to check that K_1 and K_2 satisfy the Legendre condition (1) of Theorem 4.3 in Chapter 4, p. 59. To check the Legendre conditions (2), (3) and locate the curvature spheres of μ, we compute dF and $d\eta$. We consider vectors $X \in T_x M^{n-1}$ and

$$Y = Y_0 u_n + \cdots + Y_m u_{n+m} \in T_y S^m,$$

and compute

$$dF(X, 0) = d\hat{f}(X) + (Xf_n)y, \quad d\eta(X, 0) = d\hat{\xi}(X) + (X\xi_n)y, \tag{5.27}$$

$$dF(0, Y) = f_n(x)Y, \quad d\eta(0, Y) = \xi_n(x)Y. \tag{5.28}$$

Note that when $y = u_n$, we have

$$dF(X, 0) = df(X), \qquad d\eta(X, 0) = d\xi(X).$$

The Legendre condition (3), $dF \cdot \eta = 0$, follows easily from $df \cdot \xi = 0$ and $y \cdot Y = 0$. To check condition (2), first note that $dF(X, 0)$ and $d\eta(X, 0)$ are never simultaneously zero for $X \neq 0$, since $df(X)$ and $d\xi(X)$ are never both zero for $X \neq 0$. Then equations (5.27) and (5.28) imply that $dF(X, Y)$ and $d\eta(X, Y)$ are never both zero if $X \neq 0$. On the other hand, $dF(0, Y)$ and $d\eta(0, Y)$ vanish simultaneously at (x, y) for a nonzero $Y \in T_y S^m$ if and only if $f_n(x)$ and $\xi_n(x)$ are both zero, i.e., $f(x)$ and $\xi(x)$ both lie in $\mathbf{R}^{n-1}$. This means that the line through $f(x)$ in the direction $\xi(x)$ lies in $\mathbf{R}^{n-1}$. This line is the locus of centers of the spheres in $\mathbf{R}^n$ corresponding to the points on the line $\lambda(x)$ lying on Q^{n+1}. These spheres all meet $\mathbf{R}^{n-1}$ orthogonally, as does the one plane in the parabolic pencil of spheres determined by $\lambda(x)$. Condition (2) does not automatically hold at points (x, y) of this type. This is easy to understand geometrically. If $f(x)$ and $\xi(x)$ both lie in $\mathbf{R}^{n-1}$, then they are fixed pointwise by the group $SO(m + 1)$ of isometries of $\mathbf{R}^{n+m}$ that keep $\mathbf{R}^{n-1}$ pointwise fixed. So the contact element $(f(x), \xi(x))$ is also fixed by these rotations, and the map from $M^{n-1} \times S^m$ into the space of contact elements is not an immersion at such points.

We now find the curvature spheres of μ at a point (x, y) where the Legendre condition (2) does hold. As before, the curvature spheres are determined by the condition

$$d(rF + s\eta)(X, Y) = 0.$$

From equation (5.27), we see that

$$d(rF + s\eta)(X, 0) = 0 \Leftrightarrow d(rf + s\xi)(X) = 0. \tag{5.29}$$

Thus, $(X, 0)$ is a principal vector for μ at (x, y) if and only if X is a principal vector for λ at x. The curvature sphere $[rK_1 + sK_2]$ of μ with principal vector $(X, 0)$ corresponds to the curvature sphere $[rk_1 + sk_2]$ of λ with principal vector X.

The new curvature sphere of μ is easily found from equation (5.28). For any Y in $T_y S^m$, and for any fixed x, we have

$$d(\xi_n(x)F - f_n(x)\eta)(0, Y) = 0.$$

Hence,

$$[K] = [\xi_n(x)K_1 - f_n(x)K_2] \tag{5.30}$$

is a curvature sphere of multiplicity at least m at (x, y). From equation (2.16) of Chapter 2, p. 16, we see that if $\xi_n(x) \neq 0$, then $[K]$ represents the oriented sphere with center

$$\xi_n(x)\hat{f}(x) - f_n(x)\hat{\xi}(x),$$

and signed radius $-f_n(x)/\xi_n(x)$. The center of $[K]$ is the unique point of intersection of the line through $f(x)$ in the direction $\xi(x)$ with the axis $\mathbf{R}^{n-1}$, and $[K]$ meets $\mathbf{R}^{n-1}$ orthogonally. If $\xi_n(x) = 0$, then $[K]$ represents the hyperplane through $f(x)$ with normal $\xi(x) = \hat{\xi}(x)$. This hyperplane also meets $\mathbf{R}^{n-1}$ orthogonally. Thus, in either case, $[K]$ is orthogonal to

$$u_{n+i} = e_{n+2+i}, \quad 0 \leq i \leq m,$$

and $[K]$ is contained in the $(n + 1)$-dimensional linear subspace E of $\mathbf{P}^{n+m+2}$ whose orthogonal complement is the $(m + 1)$-dimensional vector space

$$E^{\perp} = \text{Span}\{e_{n+2}, \ldots, e_{n+m+2}\}, \tag{5.31}$$

which has signature $(m + 1, 0)$.

Thus there are two possibilities for the number of distinct curvature spheres of μ at (x, y). The first case is when $[K]$ is not equal to one of the curvature spheres induced from the curvature spheres of λ, i.e., when none of the curvature spheres on the line $\lambda(x)$ are orthogonal to $\mathbf{R}^{n-1}$. In that case, $[K]$ has multiplicity m, and its curvature surface through (x, y) is $\{x\} \times S^m$, along which $[K]$ is constant. The other curvature spheres of μ at (x, y) correspond exactly to the curvature spheres of λ at x. Since λ is Dupin, these curvature spheres of μ are also constant along their curvature surfaces, which have the form $S \times \{y\}$, where S is a curvature surface of λ through x. The number $\gamma(x, y)$ of distinct curvature spheres of μ at (x, y) is $g + 1$. The second case is when $[K]$ is the curvature sphere induced from one of the curvature spheres $[k]$ of λ at x. Then $[K]$ has multiplicity $m + \nu$, where ν is the multiplicity of $[k]$. The curvature surface of $[K]$ through (x, y) is $S \times S^m$, where S is the curvature surface of $[k]$ through x. The curvature sphere $[K]$ is clearly constant along this curvature surface. In this case, the number of distinct curvature sphere $\gamma(x, y) = g$. We summarize these results in the following proposition.

Proposition 5.9. *Suppose that* $\lambda : M^{n-1} \to \Lambda^{2n-1}$ *is a proper Dupin submanifold with g distinct curvature spheres. The surface of revolution construction yields a Dupin submanifold μ defined on all of $M^{n-1} \times S^m$, except those points where the spheres in the parabolic pencil determined by the line $\lambda(x)$ are all orthogonal to the axis $\mathbf{R}^{n-1}$. For (x, y) in the domain of μ, the number $\gamma(x, y)$ of distinct curvature spheres of μ at (x, y) is as follows:*

(a) $\gamma(x, y) = g + 1$, *if none of the curvature spheres of λ at x are orthogonal to the axis $\mathbf{R}^{n-1}$.*
(b) $\gamma(x, y) = g$ *otherwise.*

Thus, as with the other constructions, there are two cases in which this construction yields a proper Dupin submanifold; either no curvature sphere of λ is ever orthogonal to the axis $\mathbf{R}^{n-1}$ or one of the curvature spheres of λ is always orthogonal to the axis.

Next we prove a proposition concerning the surface of revolution construction that will be used in the proof of Theorem 5.16 which states that a compact proper Dupin hypersurface embedded in Euclidean space with $g > 2$ distinct principal curvatures must be irreducible. That result and the following proposition were proved by Cecil, Chi, and Jensen [41].

Proposition 5.10. *Suppose that* $\mu : M^{n-1} \times S^m \to \Lambda^{2(n+m)-1}$ *is a Legendre submanifold that is obtained from a proper Dupin submanifold $\lambda : M^{n-1} \to \Lambda^{2n-1}$ by the the surface of revolution construction. If there exists a Lie sphere transformation β such that the point sphere map of $\beta\mu$ is an immersion, then there exists a Lie sphere transformation α such that the point sphere map of $\alpha\lambda$ is an immersion.*

Proof. Using the terminology of Section 4.4, suppose that α is the Lie sphere (projective) transformation induced by an orthogonal transformation $A \in O(n + 1, 2)$ of $\mathbf{R}_2^{n+3}$. Let

$$[W_1] : M^{n-1} \to Q^{n+1}$$

be the point sphere map of $\alpha\lambda$ which is determined by the equation

$$\langle W_1(x), e_{n+m+3} \rangle = 0,$$

for $x \in M^{n-1}$. The map $[W_1]$ is an immersion if and only if $[W_1(x)]$ is not a curvature sphere of $\alpha\lambda$ for any $x \in M^{n-1}$. By Theorem 4.7 in Section 4.4, the curvature spheres of $\alpha\lambda$ are of the form $\alpha[k]$, where $[k]$ is a curvature sphere of λ. Thus, the point sphere map $[W_1]$ of $\alpha\lambda$ is an immersion if and only if

$$\langle Ak(x), e_{n+m+3} \rangle \neq 0, \tag{5.32}$$

for all curvature spheres $[k(x)]$ for all $x \in M^{n-1}$. If we apply the Lie transformation α^{-1} induced by $A^{-1} \in O(n+1, 2)$, then the inequality (5.32) becomes

$$\langle k(x), A^{-1}e_{n+m+3} \rangle = \langle k(x), v \rangle \neq 0, \tag{5.33}$$

for all curvature spheres $[k(x)]$ for all $x \in M^{n-1}$, where

$$v = A^{-1}e_{n+m+3},$$

is a unit timelike vector in $\mathbf{R}_2^{n+3}$. Conversely, if there exists a unit timelike vector $v \in \mathbf{R}_2^{n+3}$ such that

$$\langle k, v \rangle \neq 0, \tag{5.34}$$

for all curvature sphere maps of λ, then if α is a Lie sphere transformation such that $\alpha(v) = e_{n+m+3}$, the point sphere map of $\alpha\lambda$ will be an immersion.

Suppose that there exists a Lie sphere transformation β such that the point sphere map of $\beta\mu$ is an immersion. Then there exists a unit timelike vector $q \in \mathbf{R}_2^{n+m+3}$ such that

$$\langle K, q \rangle \neq 0, \tag{5.35}$$

for all curvature sphere maps K of μ. We can write q in coordinates as

$$q = (q_1, q_2, \hat{q}, w, q_{n+m+3}), \tag{5.36}$$

where $\hat{q} = (q_3, \ldots, q_{n+1}) \in \mathbf{R}^{n-1}$ and $w = (q_{n+2}, \ldots, q_{n+m+2})$.

By equations (5.29) and (5.30), we know that the curvature spheres of μ are of two types. The first type is

$$[K(x, y)] = [rK_1(x, y) + sK_2(x, y)], \tag{5.37}$$

where r, s are real numbers such that

$$[k(x)] = [rk_1(x) + sk_2(x)], \tag{5.38}$$

is curvature sphere of λ at $x \in M^{n-1}$. The second type is

$$[K(x, y)] = [\xi_n(x)K_1(x, y) - f_n(x)K_2(x, y)], \tag{5.39}$$

which is the new curvature sphere introduced by the surface of revolution construction.

For a curvature sphere of the first type, as in equation (5.37), one can compute using equations (5.4)–(5.7) and (5.25)–(5.26) that

$$\langle K(x, y), q \rangle = -q_1 \left(r \left(\frac{1 + f \cdot f}{2} \right) + sf \cdot \xi \right) \tag{5.40}$$

$$+ q_2 \left(r \left(\frac{1 - f \cdot f}{2} \right) - sf \cdot \xi \right) + (r\hat{f}(x) + s\hat{\xi}(x)) \cdot \hat{q}$$

$$+ (rf_n(x) + s\xi_n(x))(y \cdot w) - sq_{n+m+3},$$

since $F \cdot F = f \cdot f$, $\eta \cdot \eta = \xi \cdot \xi$, and $F \cdot \eta = f \cdot \xi$. Note that all terms depend only on $x \in M^{n-1}$ except for the term

$$(rf_n(x) + s\xi_n(x))(y \cdot w). \tag{5.41}$$

If we take $y = e_{n+2}$ in this expression, then we get from equations (5.23)–(5.26) and (5.35) that

$$0 \neq \langle K(x, y), q \rangle = \langle k(x), v \rangle, \tag{5.42}$$

for $k(x) = rk_1(x) + sk_2(x)$ and $v = \pi(q)$, where π is orthogonal projection of $\mathbf{R}_2^{n+m+3}$ onto $\mathbf{R}_2^{n+3}$ given by

$$v = \pi \left(\sum_{i=1}^{n+m+3} q_i e_i \right) = \sum_{i=1}^{n+2} q_i e_i + q_{n+m+3} e_{n+m+3}. \tag{5.43}$$

The vector v is timelike, since q is a unit timelike vector, and

$$\langle v, v \rangle = \langle q, q \rangle - (q_{n+3}^2 + \cdots + q_{n+m+2}^2). \tag{5.44}$$

Thus, $v/|v|$ is a unit timelike vector in $\mathbf{R}_2^{n+3}$. Since $\langle K, q \rangle \neq 0$ by equation (5.35) for all curvature spheres K of μ, in particular those of type (5.37), we see from equation (5.42) that $\langle k, v/|v| \rangle \neq 0$ for all curvature sphere maps $[k]$ of λ. Thus, as discussed in equation (5.34), there exists a Lie sphere transformation α such that the point sphere map of $\alpha\lambda$ is an immersion. $\square$

5.3 Lie Sphere Geometric Criterion for Reducibility

We now find a Lie sphere geometric criterion for when a Dupin submanifold is reducible to a lower-dimensional Dupin submanifold. First, note that the totally umbilic case of a proper Dupin submanifold with one distinct curvature sphere is well known. These are all Lie equivalent to the Legendre submanifold induced by an open subset of a standard metric sphere S^{n-1} embedded in $\mathbf{R}^n$. A standard sphere can be obtained from a point by any of the standard constructions. From now on, we will only consider the case in which the number of distinct curvature spheres is greater than one.

We say that a Dupin submanifold η that is obtained from a Dupin submanifold λ by one of the standard constructions is *reducible to* λ. More generally, a Dupin submanifold μ that is Lie equivalent to such a Dupin submanifold η is also said to be *reducible to* λ.

In general, as we see from Propositions 5.2, 5.7 and 5.9, the application of one of the standard constructions to a proper Dupin submanifold with g distinct curvature spheres produces a proper Dupin submanifold with $g + 1$ distinct curvature spheres defined on an open subset of $M^{n-1} \times S^m$. (Example 5.6 shows that this is not always the case, however.) Pinkall [150, p. 438] found the following simple criterion for reducibility in this general situation.

Theorem 5.11. *A proper Dupin submanifold* $\mu : W^{d-1} \to \Lambda^{2d-1}$ *with* $g + 1 \geq 2$ *distinct curvature spheres is reducible to a proper Dupin submanifold* λ *with* g *distinct curvature spheres if and only if* μ *has a curvature sphere* $[K]$ *of multiplicity* $m \geq 1$ *that lies in a* $(d + 1 - m)$*-dimensional linear subspace of* $\mathbf{P}^{d+2}$.

Proof. Let $n = d - m$ in order to agree with the notation used in the previous section. Since μ has at least two distinct curvature spheres, we have

$$d - 1 - m \geq 1,$$

i.e., $n \geq 2$. For each of the three constructions, it was shown that if the constructed Dupin submanifold η has one more curvature sphere than the original Dupin submanifold λ, then the new curvature sphere $[K]$ has multiplicity m and lies in a $(d+1-m)$-dimensional linear subspace E of $\mathbf{P}^{d+2}$. The same holds for a Dupin submanifold μ that is Lie equivalent to such a Dupin submanifold η.

Conversely, if there exists a curvature sphere $[K]$ of multiplicity m that lies in an $(n + 1)$-dimensional space E, then the signature of $\langle , \rangle$ on the $(m + 1)$-dimensional vector space $E^\perp$ must be $(m + 1, 0)$, $(m, 1)$ or $(m, 0)$. Otherwise, $E \cap Q^{d+1}$ is either empty or consists of a single point or a line (see Corollary 2.5 of Chapter 2, p. 21). However, the curvature sphere map $[K]$ is an immersion of the $(n - 1)$-dimensional space of leaves M^{n-1} of the principal foliation corresponding to $[K]$, and its image cannot be contained in a single line.

If the signature of $\langle , \rangle$ on $E^\perp$ is $(m+1, 0)$, then there is a Lie sphere transformation α, induced by an orthogonal transformation $A \in O(d + 1, 2)$, which takes $E^\perp$ to the space in equation (5.31). For the Dupin submanifold $\eta = \alpha\mu$, the centers of the curvature spheres in the family $[AK]$ all lie in the space

$$\mathbf{R}^{n-1} = \mathrm{Span}\{e_3, \ldots, e_{n+1}\} \subset \mathbf{R}^n = \mathrm{Span}\{e_3, \ldots, e_{n+2}\}. \quad (5.45)$$

The proper Dupin submanifold η is an envelope of this family of curvature spheres $[AK]$, with each curvature sphere tangent to the envelope along a leaf of the principal foliation corresponding to $[AK]$. Since the family of curvature spheres $[AK]$ is invariant under $SO(m + 1)$, the subgroup of $SO(d)$ consisting of isometries that keep the axis $\mathbf{R}^{n-1}$ pointwise fixed, the envelope of these curvature spheres is also invariant under $SO(m + 1)$. Thus η is an open subset of a surface of revolution. The profile submanifold λ in $\mathbf{R}^n$ of this surface of revolution is locally obtained by taking those contact elements in $\mathbf{R}^n$ which are in the image of η. Each curvature surface of $[AK]$ is the orbit of a contact element in the image of λ under the action of the group $SO(m + 1)$. Since the multiplicity m of $[AK]$ is accounted for by the action of this group, the profile submanifold has one less curvature sphere than η (and hence μ) at each point.

Similarly, if the signature of $\langle , \rangle$ on $E^\perp$ is $(m, 1)$, then $E^\perp$ can be mapped by a Lie sphere transformation α induced by $A \in O(d + 1, 2)$ to the space in equation (5.15) given in the tube construction. Then each curvature sphere in the family $[AK]$ has radius $-\varepsilon$ and has center in the space $\mathbf{R}^n$ given in equation (5.45). Since the map $[AK]$ factors through an immersion of the space of leaves M^{n-1} of the principal foliation, the locus of centers of these spheres factors through an immersion f of M^{n-1} into $\mathbf{R}^n$. The proper Dupin submanifold $\eta = \alpha\mu$ is an envelope of this family of curvature spheres, and it is obtained from the Legendre submanifold λ induced from the hypersurface f in $\mathbf{R}^n$ via the tube construction. Since the multiplicity of $[AK]$ is accounted for by the tube construction, λ has one less curvature sphere than μ.

Finally, if the signature of $\langle , \rangle$ on $E^\perp$ is $(m, 0)$, then $E^\perp$ can be mapped by a Lie sphere transformation α induced by $A \in O(d + 1, 2)$ to the space in equation (5.21) in the cylinder construction. The family $[AK]$ of curvature spheres consists of hyperplanes orthogonal to the space $\mathbf{R}^n$ given in equation (5.45). The proper Dupin submanifold $\eta = \alpha\mu$ is an envelope of this family of hyperplanes, with each hyperplane tangent to the envelope along a leaf of the principal foliation. This family of hyperplanes is invariant under the action of the group H of translations of $\mathbf{R}^d$ in directions orthogonal to $\mathbf{R}^n$, and so is the envelope. Each leaf of the principal foliation is the orbit of a single contact element in $\mathbf{R}^n$ under the action of H. These contact elements in $\mathbf{R}^n$ determine the original proper Dupin submanifold λ from which η is obtained by the cylinder construction. Again, it is clear that λ has one less curvature sphere than μ at each point, since the multiplicity m of the curvature sphere $[AK]$ equals the codimension of $\mathbf{R}^n$ in $\mathbf{R}^d$. $\square$

Pinkall [150, p. 438] also formulated his local criterion for reducibility to handle the case where the number of distinct curvature spheres of μ is the same as the number of distinct curvature spheres of λ, as in Example 5.6. For this theorem, we do not take into account the multiplicity of the curvature sphere $[K]$. The result also holds for a proper Dupin submanifold with one curvature sphere at each point, so we also include that case.

The following version of the proof of Pinkall's criterion for reducibility was published in [41]. This proof makes use of the fact that a proper Dupin submanifold must be algebraic and hence analytic. Pinkall [148] sent a sketch of a proof of that fact to the author in a letter in 1984. A complete proof of the algebraicity of proper Dupin submanifolds, based on Pinkall's approach, was given in a paper of Cecil, Chi and Jensen [42].

Theorem 5.12. *A connected proper Dupin submanifold* $\mu : W^{d-1} \to \Lambda^{2d-1}$ *is reducible if and only if there exists a curvature sphere* $[K]$ *of* μ *that lies in a linear subspace of* $\mathbf{P}^{d+2}$ *of codimension at least two.*

Proof. First, assume that μ is reducible. By definition this means that for every $x \in W^{d-1}$, there exists a neighborhood of x such that the restriction of μ to this neighborhood is Lie equivalent to the end product of one of the standard constructions.

For each of these constructions it was shown that one of the curvature spheres $[K]$ lies in a space of codimension at least two in $\mathbf{P}^{d+2}$. For each $x \in W^{d-1}$, let $m_x \geq 1$ be the largest positive integer such that for some neighborhood U_x of x, the restriction of the curvature sphere map $[K]$ to U_x lies in a linear subspace of codimension $m_x + 1$ in $\mathbf{P}^{d+2}$. Choose x_0 to be a point where m_x attains its maximum value m on W^{d-1}. Then there exist linearly independent vectors $v_1, \ldots, v_{m+1}$ in $\mathbf{R}_2^{d+3}$ such that on a neighborhood U_{x_0} of x_0, we have

$$\langle K, v_i \rangle = 0, \quad 1 \leq i \leq m + 1. \tag{5.46}$$

Since μ is analytic, the curvature sphere map $[K]$ is analytic. Then since the analytic functions $\langle K, v_i \rangle$ equal zero on the open set U_{x_0} in the connected analytic manifold W^{d-1}, they must equal zero on all of W^{d-1}. Thus, equation (5.46) holds on all of W^{d-1}, and the function $m_x = m$ for all $x \in W^{d-1}$. The curvature sphere $[K]$ lies in the space E of codimension $m + 1$ determined by equation (5.46), and so $[K]$ lies in a linear space of codimension at least two in $\mathbf{P}^{d+2}$.

The proof of the converse is essentially the same as that of Theorem 5.11. Suppose that μ has a curvature sphere $[K]$ that lies in a linear subspace E of codimension $m + 1$, where $m \geq 1$. As in the proof of Theorem 5.11, there exists a Lie sphere transformation α induced by $A \in O(d+1, 2)$ such that α maps $E^{\perp}$ to the appropriate space in equations (5.15), (5.21), (5.31), as determined by the signature of $\langle, \rangle$ on $E^{\perp}$. The proof of Theorem 5.11 deals specifically with the case where $[K]$ has multiplicity m, and so the number of curvature spheres of $\eta = \alpha\mu$ is one greater than the number of curvature spheres of λ. If $[K]$ has multiplicity greater than m, then the curvature sphere $[AK]$ of η is also equal to one of the curvature spheres of η induced from a curvature sphere $[k]$ of λ, and the multiplicity of $[K]$ is $m + l$, where l is the multiplicity of $[k]$. In that case, μ and λ have the same number of curvature spheres, as in Example 5.6. The rest of the proof is quite similar to the proof of Theorem 5.11. □

Remark 5.13. When Pinkall introduced his constructions, he also listed the following construction. Begin with a proper Dupin submanifold λ induced by an embedded proper Dupin hypersurface $M^{n-1} \subset S^n \subset \mathbf{R}^{n+1}$. The new Dupin submanifold μ is the Legendre submanifold induced from the cone C^n over M^{n-1} in $\mathbf{R}^{n+1}$ with vertex at the origin. Theorem 5.12 shows that this construction is locally Lie equivalent to the tube construction as follows. The tube construction is characterized by the fact that one curvature sphere map $[K]$ lies in a d-dimensional linear subspace E of $\mathbf{P}^{d+2}$, whose orthogonal complement has signature $(1, 1)$. Geometrically, this means that after a suitable Lie sphere transformation, all the spheres in the family $[K]$ have the same radius and their centers lie in a subspace $\mathbf{R}^{d-1} \subset \mathbf{R}^d$. For the cone construction, the new family $[K]$ of curvature spheres consists of hyperplanes through the origin (corresponding to the point $[e_1 + e_2]$ in Lie sphere geometry) that are tangent to the cone along the rulings. Since the hyperplanes also all pass through the improper point $[e_1 - e_2]$, they correspond to points in the linear subspace E, whose orthogonal complement is as follows:

$$E^{\perp} = \text{Span}\{e_1 + e_2, e_1 - e_2\}.$$

Since $E^\perp$ is spanned by e_1 and e_2, it has signature $(1, 1)$. Thus, the cone construction is Lie equivalent to the tube construction. Finally, there is one more geometric interpretation of the tube construction. Note that a family $[K]$ of curvature spheres that lies in a linear subspace whose orthogonal complement has signature $(1, 1)$ can also be considered to consist of spheres in S^d of constant radius in the spherical metric whose centers lie in a hyperplane. The corresponding proper Dupin submanifold can thus be considered to be a tube in the spherical metric over a lower-dimensional submanifold in S^d.

Remark 5.14. In a recent paper [55], Dajczer, Florit and Tojeiro studied reducibility in the context of Riemannian geometry. They formulated a concept of weak reducibility for proper Dupin submanifolds that have a flat normal bundle, including proper Dupin hypersurfaces. For hypersurfaces, their definition can be formulated as follows. A proper Dupin hypersurface $f : M^{n-1} \to \mathbf{R}^n$ (or S^n) is said to be *weakly reducible* if, for some principal curvature κ_i with corresponding principal space T_i, the orthogonal complement $T_i^\perp$ is integrable. Dajczer, Florit and Tojeiro show that if a proper Dupin hypersurface $f : M^{n-1} \to \mathbf{R}^n$ is Lie equivalent to a proper Dupin hypersurface with $g+1$ distinct principal curvatures that is obtained via one of the standard constructions from a proper Dupin hypersurface with g distinct principal curvatures, then f is weakly reducible. Thus, reducible implies weakly reducible for such hypersurfaces.

However, one can show that the open set U of the tube W^4 over V^2 in S^5 in Example 5.6 on which there are three principal curvatures at each point is reducible but not weakly reducible, because none of the orthogonal complements of the principal spaces is integrable. Of course, U is not constructed from a proper Dupin submanifold with two curvature spheres, but rather one with three curvature spheres.

In two papers by Cecil and Jensen [44]–[45], the notion of local irreducibility was used. Specifically, a proper Dupin submanifold $\lambda : M^{n-1} \to \Lambda^{2n-1}$ is said to be *locally irreducible* if there does not exist any open subset $U \subset M^{n-1}$ such that the restriction of λ to U is reducible. Theoretically, this is a stronger condition than irreducibility of λ itself. However, using the analyticity of proper Dupin submanifolds, Cecil, Chi and Jensen [41] proved the following proposition which shows that the concepts of local irreduciblity and irreducibility are equivalent.

Proposition 5.15. *Let* $\lambda : M^{n-1} \to \Lambda^{2n-1}$ *be a connected, proper Dupin submanifold. If the restriction of* λ *to an open subset* $U \subset M^{n-1}$ *is reducible, then* λ *is reducible. Thus, a connected proper Dupin submanifold is locally irreducible if and only if it is irreducible.*

Proof. Suppose there exists an open subset $U \subset M^{n-1}$ on which the restriction of λ is reducible. By Theorem 5.12 there exists a curvature sphere $[K]$ of λ and two linearly independent vectors v_1 and v_2, such that

$$\langle K(x), v_i \rangle = 0, \quad i = 1, 2,$$

for all $x \in U$. Since the curvature sphere map $[K]$ is analytic on M^{n-1}, the functions $\langle K, v_i \rangle$ are analytic on M^{n-1} for $i = 1, 2$. Since these functions are identically equal to zero on the open set U, they must equal zero on all of the connected analytic manifold M^{n-1}. Therefore, by Theorem 5.12, the proper Dupin submanifold λ : $M^{n-1} \to \Lambda^{2n-1}$ is reducible. Thus, if λ is irreducible, then it cannot have a reducible open subset, so it must be locally irreducible. $\square$

The considerations above are all of a local nature. We now want to consider the global question of when a compact proper Dupin hypersurface embedded in $\mathbf{R}^d$ or S^d is irreducible. Thorbergsson [190] showed that a compact, connected proper Dupin hypersurface immersed in $\mathbf{R}^d$ or S^d is taut, and therefore it is embedded (see Carter–West [21] or Cecil–Ryan [52, p. 121]). The following theorem was proved by Cecil, Chi and Jensen [41].

Theorem 5.16. *Let W^{d-1} be a compact, connected proper Dupin hypersurface immersed in $\mathbf{R}^d$ with $g > 2$ distinct principal curvatures. Then W^{d-1} is irreducible. That is, the Legendre submanifold induced by the hypersurface W^{d-1} is irreducible.*

Proof. As noted above, tautness implies that an immersed compact, connected proper Dupin hypersurface is embedded in $\mathbf{R}^d$. We will assume that $W^{d-1} \subset \mathbf{R}^d$ is reducible and obtain a contradiction. Let $\mu : W^{d-1} \to \Lambda^{2d-1}$ be the Legendre submanifold induced by the embedded hypersurface $W^{d-1} \subset \mathbf{R}^d$. By the proof of Theorem 5.12, the fact that μ is reducible implies that μ is equivalent by a Lie sphere transformation α to a proper Dupin submanifold $\eta = \alpha\mu : W^{d-1} \to \Lambda^{2d-1}$ that is obtained from a lower-dimensional proper Dupin submanifold $\lambda : M^{n-1} \to \Lambda^{2n-1}$ by one of the three standard constructions. Thus, W^{d-1} is diffeomorphic to $M^{n-1} \times S^m$, where $m = d - n$, and M^{n-1} must be compact, since W^{d-1} is compact. By hypothesis, μ has $g > 2$ distinct curvature spheres at each point, and thus so does η. For η obtained from λ by the tube or cylinder constructions, Propositions 5.2, 5.4, and 5.7 show that there always exist points on $M^{n-1} \times S^m$ at which the number of distinct curvature spheres is two, and therefore η cannot be obtained via the tube or cylinder constructions.

Therefore the only remaining possibility is that η is obtained from λ by the surface of revolution construction. Proposition 5.9 shows that for a surface of revolution, the number j of distinct curvature spheres on M^{n-1} must be $g - 1$ or g. Note that the sum β of the $\mathbf{Z}_2$-Betti numbers of W^{d-1} and M^{n-1} are related by the equation,

$$\beta(W^{d-1}) = \beta(M^{n-1} \times S^m) = 2\beta(M^{n-1}). \tag{5.47}$$

On the other hand, Thorbergsson showed that for a connected, compact proper Dupin hypersurface embedded in S^d, β is equal to twice the number of distinct curvature spheres. Thus, we have $\beta(W^{d-1}) = 2g$.

We know that η is Lie equivalent to μ and the point sphere map of μ is an immersion. Furthermore, η is obtained from λ by the surface of revolution construction. Thus, by Proposition 5.10, we conclude that there exists a Lie sphere transformation γ such that the point sphere map of $\gamma\lambda$ is an immersion. This point sphere map of $\gamma\lambda$ gives rise to a Euclidean projection,

$$f : M^{n-1} \to \mathbf{R}^n,$$

that is an immersed (and thus embedded) proper Dupin hypersurface. Thus, by Thorbergsson's theorem, we have $\beta(M^{n-1}) = 2j$, where j equals $g - 1$ or g. This fact, together with equation (5.47), implies that it is impossible for W^{d-1} and M^{n-1} to have the same number of distinct curvature spheres, and so M^{n-1} has $g - 1$ distinct curvature spheres. Hence, we have

$$\beta(W^{d-1}) = 2g, \qquad \beta(M^{n-1}) = 2(g - 1) = 2g - 2. \qquad (5.48)$$

Combining equations (5.47) and (5.48), we get

$$2g = 2(2g - 2) = 4g - 4,$$

and thus $g = 2$, contradicting the assumption that $g > 2$. Therefore, μ cannot be reducible. $\qquad\qquad\square$

A hypersurface in S^d is conformally equivalent to its image in $\mathbf{R}^d$ under stereographic projection. Furthermore, the proper Dupin condition is preserved under stereographic projection. Thus, as a corollary of Theorem 5.16, we conclude that a compact, connected isoparametric hypersurface in S^d is irreducible as a Dupin hypersurface if the number g of distinct principal curvatures is greater than two. This was proved earlier by Pinkall in his dissertation [146]. Of course, compactness is not really a restriction for an isoparametric hypersurface, since Münzner [123] has shown that any connected isoparametric hypersurface is contained in a unique compact, connected isoparametric hypersurface. The same is not true for proper Dupin hypersurfaces, since the completion of a proper Dupin hypersurface may not be proper Dupin. Consider, for example, the tube M^3 over a torus $T^2 \subset \mathbf{R}^3 \subset \mathbf{R}^4$ in Example 4.10, p. 69. The tube M^3 is the completion of the open subset U of M^3 on which there are three distinct principal curvatures of multiplicity one. The set U is a proper Dupin hypersurface (with two connected components), but M^3 is only Dupin, but not proper Dupin. This phenomenon is also made clear by Propositions 5.2, 5.4, 5.7, and 5.9.

There is one other geometric consequence about isoparametric hypersurfaces that is implied by the theorem. Münzner showed that an isoparametric hypersurface $M^{n-1} \subset S^n \subset \mathbf{R}^{n+1}$ is a tube of constant radius in S^n over each of its two focal submanifolds. If $g = 2$, then the isoparametric hypersurface M^{n-1} must be a standard product of two spheres,

$$S^k(r) \times S^{n-k-1}(s) \subset S^n, \quad r^2 + s^2 = 1,$$

and the two focal submanifolds are both totally geodesic spheres, $S^k(1) \times \{0\}$ and $\{0\} \times S^{n-k-1}(1)$. The isoparametric hypersurface M^{n-1} is reducible in two ways, since it can be obtained as a tube of constant radius over each of these focal submanifolds, which are not substantial in $\mathbf{R}^{n+1}$. On the other hand, if an isoparametric hypersurface M^{n-1} has $g \geq 3$ distinct principal curvatures, then each of its focal submanifolds must be substantial in $\mathbf{R}^{n+1}$. Otherwise, M^{n-1} would be reducible to such a nonsubstantial focal submanifold by the tube construction, contradicting Theorem 5.16.

Finally, Theorem 5.16 has the following corollary concerning the location of curvature spheres in $\mathbf{R}^n$.

Corollary 5.17. *Let $M^{n-1} \subset \mathbf{R}^n$ be a compact, connected proper Dupin hypersurface having $g \geq 2$ distinct curvature spheres at each point. Let $\mathbf{R}^{n-1}$ be any hyperplane in $\mathbf{R}^n$ which is disjoint from M^{n-1}. Then there must exist a curvature sphere at some point of M^{n-1} that is orthogonal to $\mathbf{R}^{n-1}$.*

Proof. Consider $\mathbf{R}^n$ embedded as a subspace of a Euclidean space $\mathbf{R}^d, d > n$. Since M^{n-1} is disjoint from $\mathbf{R}^{n-1}$, the hypersurface W^{d-1} obtained by revolving M^{n-1} about $\mathbf{R}^{n-1}$ is embedded in $\mathbf{R}^d$. If no curvature sphere of M^{n-1} intersects $\mathbf{R}^{n-1}$ orthogonally, then by Proposition 5.9, W^{d-1} is a compact reducible proper Dupin hypersurface with more than two distinct curvature spheres, contradicting Theorem 5.16. □

Geometrically, this corollary means that either the focal set of M^{n-1} in $\mathbf{R}^n$ must intersect $\mathbf{R}^{n-1}$, or there must be a point $x \in M^{n-1}$ where the tangent hyperplane to M^{n-1} is a curvature sphere to $\mathbf{R}^{n-1}$. In that case, the corresponding focal point on the normal line to M^{n-1} at x intersects $\mathbf{R}^{n-1}$ at a point on the hyperplane at infinity in the projective space $\mathbf{P}^n$ determined by the affine space $\mathbf{R}^n$. In this sense, we can say that the focal set of M^{n-1} must intersect every hyperplane that is disjoint from M^{n-1}.

5.4 Cyclides of Dupin

A proper Dupin submanifold with two distinct curvature spheres of respective multiplicities p and q is called a *cyclide of Dupin of characteristic* (p, q). These are the simplest Dupin submanifolds after the spheres, and they were first studied in $\mathbf{R}^3$ by Dupin [64] in 1822. An example of a cyclide of Dupin of characteristic $(1, 1)$ in $\mathbf{R}^3$ is a torus of revolution. The cyclides were studied by many prominent mathematicians in the nineteenth century, including Liouville [107], Cayley [29], and Maxwell [109], whose paper contains stereoscopic figures of the various types of cyclides. The long history of the classical cyclides of Dupin is given in Lilienthal [106]. (See also Banchoff [6], Cecil [31], Klein [94, pp. 56–58], Darboux [56, Vol. 2, pp. 267–269], Blaschke [10, p. 238], Eisenhart [66, pp. 312–314], Hilbert and Cohn–Vossen [90, pp. 217–219], Fladt and Baur [77, pp. 354–379], and Cecil and Ryan [50], [52, pp. 151–166], for more on the classical cyclides.) For a consideration of the cyclides in the context of computer graphics, see Degen [57], Pratt [155]–[156], Srinivas and Dutta [172]–[175], and Schrott and Odehnal [168].

For cyclides of Dupin in $\mathbf{R}^3$, it was known in the nineteenth century that every connected Dupin cyclide is Möbius equivalent to an open subset of a surface of revolution obtained by revolving a profile circle $S^1 \subset \mathbf{R}^2$ about an axis $\mathbf{R}^1 \subset \mathbf{R}^2 \subset \mathbf{R}^3$. The profile circle is allowed to intersect the axis, thereby introducing Euclidean singularities. However, the corresponding Legendre map into the space of contact elements in $\mathbf{R}^3$ is an immersion, as discussed in Section 4.3.

Higher-dimensional cyclides of Dupin appeared in the study of isoparametric hypersurfaces in spheres. Cartan knew that an isoparametric hypersurface in a sphere with two curvature spheres must be a standard product of spheres,

$$S^p(r) \times S^q(s) \subset S^n(1) \subset \mathbf{R}^{p+1} \times \mathbf{R}^{q+1} = \mathbf{R}^{p+q+2}, \quad r^2 + s^2 = 1.$$

Cecil and Ryan [47] showed that a compact proper Dupin hypersurface M^{n-1} embedded in S^n with two distinct curvature spheres must be Möbius equivalent to a standard product of spheres. The proof, however, uses the compactness of M^{n-1} in an essential way. Later, Pinkall [150] used Lie sphere geometric techniques to obtain a local classification of the higher-dimensional cyclides of Dupin that is analogous to the classical result. In this section, we will prove Pinkall's theorem and then derive a local Möbius geometric classification from it. Pinkall's result is the following.

Theorem 5.18.
(a) *Every connected cyclide of Dupin is contained in a unique compact, connected cyclide of Dupin.*
(b) *Any two cyclides of Dupin of the same characteristic are locally Lie equivalent.*

Before proving the theorem, we consider some models for compact cyclides of Dupin. The results of Section 5.2 show that one can obtain a cyclide of Dupin of characteristic (p, q) by applying any of the standard constructions (tube, cylinder or surface of revolution) to a p-sphere $S^p \subset \mathbf{R}^{p+1} \subset \mathbf{R}^n$, where $n = p+q+1$. Another simple model of a cyclide of Dupin is obtained by considering the Legendre submanifold induced by a totally geodesic $S^q \subset S^n$, as a submanifold of codimension $p+1$. Such a sphere is one of the two focal submanifolds of the family of isoparametric hypersurfaces obtained by taking tubes over S^q in S^n. The other focal submanifold is a totally geodesic p-sphere S^p in S^n. We now explicitly parametrize this Legendre submanifold by k_1 and k_2 satisfying the conditions (1)–(3) of Theorem 4.3 in Chapter 4, p. 59. Let

$$\{e_1, \ldots, e_{n+3}\}$$

be the standard orthonormal basis for $\mathbf{R}_2^{n+3}$, and let

$$\Omega = \mathrm{Span}\{e_1, \ldots, e_{q+2}\}, \qquad \Omega^\perp = \mathrm{Span}\{e_{q+3}, \ldots, e_{n+3}\}. \tag{5.49}$$

These spaces have signatures $(q+1, 1)$ and $(p+1, 1)$, respectively. The intersection $\Omega \cap Q^{n+1}$ is given in homogeneous coordinates by

$$x_1^2 = x_2^2 + \cdots + x_{q+2}^2, \qquad x_{q+3} = \cdots = x_{n+3} = 0.$$

This set is diffeomorphic to the unit sphere S^q in

$$\mathbf{R}^{q+1} = \mathrm{Span}\{e_2, \ldots, e_{q+2}\},$$

by the diffeomorphism $\phi : S^q \to \Omega \cap Q^{n+1}$, $\phi(v) = [e_1 + v]$. Similarly, $\Omega^\perp \cap Q^{n+1}$ is diffeomorphic to the unit sphere S^p in

$$\mathbf{R}^{p+1} = \mathrm{Span}\{e_{q+3}, \ldots, e_{n+2}\}$$

by the diffeomorphism $\psi : S^p \to \Omega^\perp \cap Q^{n+1}$, $\psi(u) = [u + e_{n+3}]$. The Legendre submanifold $\lambda : S^p \times S^q \to \Lambda^{2n-1}$ is defined by

$$\lambda(u, v) = [k_1, k_2] \quad \text{with } [k_1(u, v)] = [\phi(v)], \quad [k_2(u, v)] = [\psi(u)]. \quad (5.50)$$

It is easy to check that the Legendre conditions (1)–(3) are satisfied by the pair $\{k_1, k_2\}$. To find the curvature spheres of λ, we decompose the tangent space to $S^p \times S^q$ at a point (u, v) as

$$T_{(u,v)} S^p \times S^q = T_u S^p \times T_v S^q.$$

Then $dk_1(X, 0) = 0$ for all $X \in T_u S^p$, and $dk_2(Y) = 0$ for all Y in $T_v S^q$. Thus, $[k_1]$ and $[k_2]$ are curvature spheres of λ with respective multiplicities p and q. Furthermore, the image of $[k_1]$ lies in the set $\Omega \cap Q^{n+1}$, and the image of $[k_2]$ is contained in $\Omega^\perp \cap Q^{n+1}$. The essence of Pinkall's proof is to show that this type of relationship always holds between the two curvature spheres of a cyclide of Dupin.

Proof of Theorem 5.18. Suppose that $\lambda : M^{n-1} \to \Lambda^{2n-1}$ is a connected cyclide of Dupin of characteristic (p, q) with $p + q = n - 1$. We may take $\lambda = [k_1, k_2]$, where $[k_1]$ and $[k_2]$ are the curvature spheres with respective multiplicities p and q. Each curvature sphere map factors through an immersion of the space of leaves of its principal foliation. Thus, locally on M^{n-1}, we can take a principal coordinate system (u, v) defined on an open set

$$W = U \times V \subset \mathbf{R}^p \times \mathbf{R}^q,$$

such that

(i) $[k_1]$ depends only on v, and $[k_2]$ depends only on u, for all $(u, v) \in W$.
(ii) $[k_1(W)]$ and $[k_2(W)]$ are submanifolds of Q^{n+1} of dimensions q and p, respectively.

Note that, in general, such a principal coordinate system cannot be found in the case of a proper Dupin submanifold with $g > 2$ curvature spheres (see Cecil–Ryan [52, p. 182]).

Now let (u, v) and $(\bar{u}, \bar{v})$ be any two points in W. From (i), we have

$$\langle k_1(u, v), k_2(\bar{u}, \bar{v}) \rangle = \langle k_1(v), k_2(\bar{u}) \rangle = \langle k_1(\bar{u}, v), k_2(\bar{u}, v) \rangle = 0. \quad (5.51)$$

Let E be the smallest linear subspace of $\mathbf{P}^{n+2}$ containing the q-dimensional submanifold $[k_1(W)]$. By equation (5.51), we have

$$[k_1(W)] \subset E \cap Q^{n+1}, \quad [k_2(W)] \subset E^\perp \cap Q^{n+1}. \quad (5.52)$$

The dimensions of E and $E^\perp$ as subspaces of $\mathbf{P}^{n+2}$ satisfy

$$\dim E + \dim E^\perp = n + 1 = p + q + 2. \quad (5.53)$$

We claim that $\dim E = q + 1$ and $\dim E^\perp = p + 1$. To see this, suppose first that $\dim E > q + 1$. Then $\dim E^\perp \leq p$, and $E^\perp \cap Q^{n+1}$ cannot contain the p-dimensional submanifold $k_2(W)$. Similarly, assuming that $\dim E^\perp > p + 1$ leads to a contradiction. Thus we have

$$\dim E \leq q + 1, \qquad \dim E^{\perp} \leq p + 1.$$

This and equation (5.53) imply that dim $E = q+1$ and dim $E^{\perp} = p+1$. Furthermore, from the fact that $E \cap Q^{n+1}$ and $E^{\perp} \cap Q^{n+1}$ contain submanifolds of dimensions q and p, respectively, it is easy to deduce that the Lie inner product $\langle , \rangle$ has signature $(q + 1, 1)$ on E and $(p + 1, 1)$ on $E^{\perp}$. Then since $E \cap Q^{n+1}$ and $E^{\perp} \cap Q^{n+1}$ are diffeomorphic to S^q and S^p, respectively, the inclusions in equation (5.52) are open subsets. If A is a Lie sphere transformation that takes E to the space Ω in equation (5.49), and thus takes $E^{\perp}$ to $\Omega^{\perp}$, then $A\lambda(W)$ is an open subset of the standard model in equation (5.50). Both assertions in Theorem 5.18 are now clear. $\square$

We now turn to the Möbius geometric classification of the cyclides of Dupin. For the classical cyclides in $\mathbf{R}^3$, this was known in the nineteenth century. K. Voss [193] announced the classification in Theorem 5.19 below for the higher-dimensional cyclides, but he did not publish a proof. The theorem follows quite directly from Theorem 5.18 and the results of the previous section on surfaces of revolution. The theorem is phrased in terms embedded hypersurfaces in $\mathbf{R}^n$. Thus we are excluding the standard model given in equation (5.50), where the Euclidean projection is not an immersion. Of course, the Euclidean projection of a parallel submanifold to the standard model is an embedding. The following proof was also given in [34].

Theorem 5.19.
(a) *Every connected cyclide of Dupin M^{n-1} of characteristic (p, q) embedded in $\mathbf{R}^n$ is Möbius equivalent to an open subset of a hypersurface of revolution obtained by revolving a q-sphere $S^q \subset \mathbf{R}^{q+1} \subset \mathbf{R}^n$ about an axis of revolution $\mathbf{R}^q \subset \mathbf{R}^{q+1}$ or a p-sphere $S^p \subset \mathbf{R}^{p+1} \subset \mathbf{R}^n$ about an axis $\mathbf{R}^p \subset \mathbf{R}^{p+1}$.*
(b) *Two such hypersurfaces are Möbius equivalent if and only if they have the same value of $\rho = |r|/a$, where r is the signed radius of the profile sphere S^q and $a > 0$ is the distance from the center of S^q to the axis of revolution.*

Proof. We always work with the Legendre submanifold induced by the embedding of M^{n-1} into $\mathbf{R}^n$. By Theorem 5.18, every connected cyclide is contained in a unique compact, connected cyclide. Thus, it suffices to classify compact, connected cyclides up to Möbius equivalence. Consider a compact, connected cyclide

$$\lambda : S^p \times S^q \to \Lambda^{2n-1}, \qquad p + q = n - 1,$$

of characteristic (p, q). By Theorem 5.18, there is a linear space E of $\mathbf{P}^{n+2}$ with signature $(q + 1, 1)$ such that the two curvature sphere maps,

$$[k_1] : S^q \to E \cap Q^{n+1}, \qquad [k_2] : S^p \to E^{\perp} \cap Q^{n+1},$$

are diffeomorphisms.

Möbius transformations are precisely those Lie sphere transformations A satisfying $A[e_{n+3}] = [e_{n+3}]$. Thus we decompose e_{n+3} as

$$e_{n+3} = \alpha + \beta, \qquad \alpha \in E, \quad \beta \in E^{\perp}. \tag{5.54}$$

Note that since $\langle \alpha, \beta \rangle = 0$, we have

$$-1 = \langle e_{n+3}, e_{n+3} \rangle = \langle \alpha, \alpha \rangle + \langle \beta, \beta \rangle.$$

Hence, at least one of the two vectors α, β is timelike. First, suppose that β is timelike. Let Z be the orthogonal complement of β in $E^{\perp}$. Then Z is a $(p + 1)$-dimensional vector space on which the restriction of $\langle , \rangle$ has signature $(p + 1, 0)$. Since $Z \subset e_{n+3}^{\perp}$, there is a Möbius transformation A such that

$$A(Z) = S = \mathrm{Span}\{e_{q+3}, \ldots, e_{n+2}\}.$$

The curvature sphere map $[Ak_1]$ of the Dupin submanifold $A\lambda$ is a q-dimensional submanifold in the space $S^{\perp} \cap Q^{n+1}$. By equation (2.14) of Chapter 2, p. 16, this means that these spheres all have their centers in the space

$$\mathbf{R}^q = \mathrm{Span}\{e_3, \ldots, e_{q+2}\}.$$

Note that

$$\mathbf{R}^q \subset \mathbf{R}^{q+1} = \mathrm{Span}\{e_3, \ldots, e_{q+3}\} \subset \mathbf{R}^n = \mathrm{Span}\{e_3, \ldots, e_{n+2}\}.$$

As we see from the proof of Theorem 5.11, this means that the Dupin submanifold $A\lambda$ is a hypersurface of revolution in $\mathbf{R}^n$ obtained by revolving a q-dimensional profile submanifold in $\mathbf{R}^{q+1}$ about the axis $\mathbf{R}^q$. Moreover, since $A\lambda$ has two distinct curvature spheres, the profile submanifold has only one curvature sphere. Thus, it is an umbilical submanifold of $\mathbf{R}^{q+1}$.

Four cases are naturally distinguished by the nature of the vector α in equation (5.54). Geometrically, these correspond to different singularity sets of the Euclidean projection of $A\lambda$. Such singularities correspond exactly with the singularities of the Euclidean projection of λ, since the Möbius transformation A preserves the rank of the Euclidean projection. Since we have assumed that β is timelike, we know that for all $u \in S^p$,

$$\langle k_2(u), e_{n+3} \rangle = \langle k_2(u), \alpha + \beta \rangle = \langle k_2(u), \beta \rangle \neq 0,$$

because the orthogonal complement of β in $E^{\perp}$ is spacelike. Thus, the curvature sphere $[Ak_2]$ is never a point sphere. However, it is possible for $[Ak_1]$ to be a point sphere.

CASE 1. $\alpha = 0$.

In this case, the curvature sphere $[Ak_1]$ is a point sphere for every point in $S^p \times S^q$. The image of the Euclidean projection of $A\lambda$ is precisely the axis $\mathbf{R}^q$. The cyclide $A\lambda$ is the Legendre submanifold induced from $\mathbf{R}^q$ as a submanifold of codimension $p + 1$ in $\mathbf{R}^n$. This is, in fact, the standard model of equation (5.50). However, since the Euclidean projection is not an immersion, this case does not lead to any of the embedded hypersurfaces classified in part (a) of the theorem.

In the remaining cases, we can always arrange that the umbilic profile submanifold is a q-sphere and not a q-plane. This can be accomplished by first inverting $\mathbf{R}^{q+1}$ in a sphere centered at a point on the axis $\mathbf{R}^q$ which is not on the profile submanifold, if necessary. Such an inversion preserves the axis of revolution $\mathbf{R}^q$. After a Euclidean translation, we may assume that the center of the profile sphere is a point $(0, a)$ on the x_{q+3}-axis ℓ in $\mathbf{R}^{q+1}$, as in Figure 5.1. The center of the profile sphere cannot lie on the axis of revolution $\mathbf{R}^q$, for then the hypersurface of revolution would be an $(n-1)$-sphere and not a cyclide of Dupin. Thus, we may take $a > 0$.

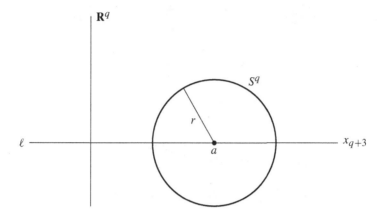

Fig. 5.1. Profile sphere S^q for the surface of revolution.

The map $[Ak_1]$ is the curvature sphere map that results from the surface of revolution construction. The other curvature sphere of $A\lambda$ corresponds exactly to the curvature sphere of the profile sphere, i.e., to the profile sphere itself. This means that the signed radius r of the profile sphere is equal to the signed radius of the curvature sphere $[Ak_2]$. Since $[Ak_2]$ is never a point sphere, we conclude that $r \neq 0$. From now on, we will identify the profile sphere with the second factor S^q in the domain of λ.

CASE 2. α *is timelike.*

In this case, for all $v \in S^q$, we have

$$\langle k_1(v), e_{n+3} \rangle = \langle k_1(v), \alpha \rangle \neq 0,$$

since the orthogonal complement of α in E is spacelike. Thus the Euclidean projection of $A\lambda$ is an immersion at all points. This corresponds to the case $|r| < a$, when the profile sphere is disjoint from the axis of revolution. Note that by interchanging the roles of α and β, we can find a Möbius transformation that takes λ to the Legendre submanifold obtained by revolving a p-sphere around an axis $\mathbf{R}^p \subset \mathbf{R}^{p+1} \subset \mathbf{R}^n$.

In the classical situation of surfaces in $\mathbf{R}^3$, the Euclidean projection of $A\lambda$ in this case is a torus of revolution (see Figure 5.2).

The Euclidean projection of λ itself is a *ring cyclide* (see Figure 5.3) if the Möbius projection of λ does not contain the improper point, or a *parabolic ring cyclide* (see

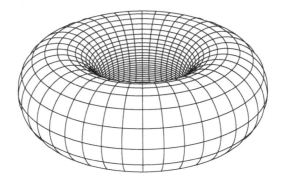

Fig. 5.2. Torus of revolution.

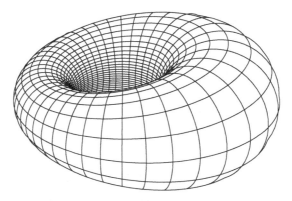

Fig. 5.3. Ring cyclide.

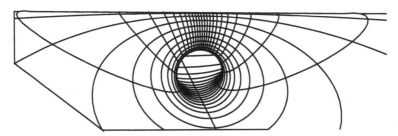

Fig. 5.4. Parabolic ring cyclide.

Figure 5.4) if the Möbius projection of λ does contain the improper point. In either case, the focal set in $\mathbf{R}^3$ consists of a pair of *focal conics*.

For a ring cyclide, the focal set consists of an ellipse and a hyperbola in mutually orthogonal planes such that the vertices of the ellipse are the foci of the hyperbola and vice-versa. For a parabolic ring cyclide, the focal set consists of two parabolas in orthogonal planes such that the vertex of each is the focus of the other. For the torus itself, the focal set consists of the core circle and the axis of revolution covered twice. This is a special case of a pair of focal conics. These classical cyclides of Dupin are discussed in more detail in the book of Cecil–Ryan [52, pp. 151–166].

CASE 3. *α is lightlike, but not zero.*

Then there is exactly one $v \in S^q$ such that

$$\langle k_1(v), e_{n+3} \rangle = \langle k_1(v), \alpha \rangle = 0. \tag{5.55}$$

This corresponds to the case $|r| = a$, where the profile sphere intersects the axis in one point. Thus, $S^p \times \{v\}$ is the set of points in $S^p \times S^q$ where the Euclidean projection is singular.

In the classical situation of surfaces in $\mathbf{R}^3$, the Euclidean projection of $A\lambda$ is a *limit torus* (see Figure 5.5), and the Euclidean projection of λ itself is a *limit spindle cyclide* (see Figure 5.6) or a *limit horn cyclide* (see Figure 5.7), if the Möbius projection of λ does not contain the improper point.

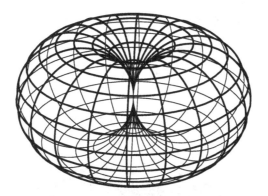

Fig. 5.5. Limit torus.

In the case where the Möbius projection of λ contains the improper point, the Euclidean projection of λ is either a *limit parabolic horn cyclide* (see Figure 5.9) or a circular cylinder (in the case where the singularity is at the improper point). For all of these surfaces except the cylinder, the focal set consists of a pair of focal conics, as in the previous case. For the cylinder, the Euclidean focal set consists only of the axis of revolution, since one of the principal curvatures is identically zero, and so the corresponding focal points are all at infinity. In Lie sphere geometry, both curvature sphere maps are plane curves on the Lie quadric, as shown in the proof of Theorem 5.18.

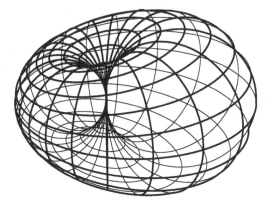

Fig. 5.6. Limit spindle cyclide.

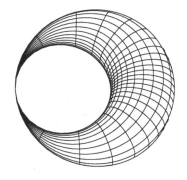

Fig. 5.7. Limit horn cyclide.

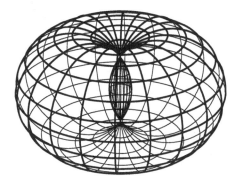

Fig. 5.8. Spindle torus.

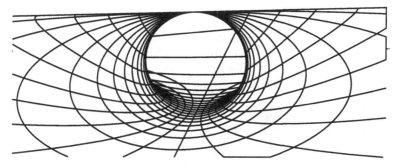

Fig. 5.9. Limit parabolic horn cyclide.

CASE 4. *α is spacelike.*

Then the condition (5.55) holds for points v in a $(q-1)$-sphere $S^{q-1} \subset S^q$. For points in $S^p \times S^{q-1}$, the point sphere map is a curvature sphere, and thus the Euclidean projection is singular. Geometrically, this is the case $|r| > a$, where the profile sphere intersects the axis $\mathbf{R}^q$ in a $(q-1)$-sphere.

In the classical situation of surfaces in $\mathbf{R}^3$, the Euclidean projection of $A\lambda$ is a *spindle torus* (see Figure 5.8), and the Euclidean projection of λ itself is a *spindle cyclide* (see Figure 5.10) or a *horn cyclide* (see Figure 5.11), if the Möbius projection does not contain the improper point.

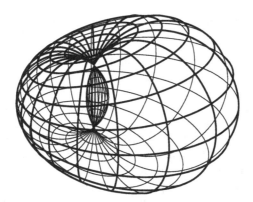

Fig. 5.10. Spindle cyclide.

In the case where the Möbius projection of λ contains the improper point, the Euclidean projection of λ is either a *parabolic horn cyclide* (see Figure 5.12) or circular cone (in the case where one of the singularities is at the improper point). For all of these surfaces except the cone, the focal set consists of a pair of focal conics. For the cone, the Euclidean focal set consists of only the axis of revolution (minus the origin), since one principal curvature is identically zero.

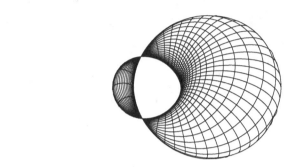

Fig. 5.11. Horn cyclide.

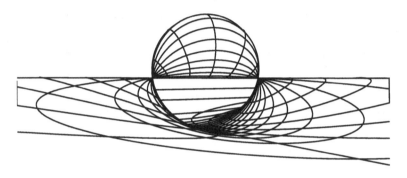

Fig. 5.12. Parabolic horn cyclide.

Of course, there are also four cases to handle under the assumption that α, instead of β, is timelike. Then the axis will be a subspace $\mathbf{R}^p \subset \mathbf{R}^{p+1}$, and the profile submanifold will be a p-sphere. The roles of p and q in determining the dimension of the singularity set of the Euclidean projection will be reversed. So if $p \neq q$, then only a ring cyclide can be represented as a hypersurface of revolution of both a q-sphere and a p-sphere. This completes the proof of part (a).

To prove part (b), we may assume that the profile sphere S^q of the hypersurface of revolution has center $(0, a)$ with $a > 0$ on the x_{q+3}-axis ℓ. Möbius classification clearly does not depend on the sign of the radius of S^q, since the two hypersurfaces of revolution obtained by revolving spheres with the same center and opposite radii differ only by the change of orientation transformation Γ (see Remark 3.4 of Chapter 3, p. 27). We now show that the ratio $\rho = |r|/a$ is invariant under the subgroup of Möbius transformations of the profile space $\mathbf{R}^{q+1}$ which take one such hypersurface of revolution to another. First, note that symmetry implies that a transformation T in this subgroup must take the axis of revolution $\mathbf{R}^q$ to itself and the axis of symmetry ℓ to itself. Since $\mathbf{R}^q$ and ℓ intersect only at 0 and the improper point ∞, the transformation T maps the set $\{0, \infty\}$ to itself. If T maps 0 to ∞, then the composition ΦT, where Φ is an inversion in a sphere centered at 0, is a member of the subgroup of transformations that map ∞ to ∞ and map 0 to 0. By Theorem 3.16 of Chapter 3, p. 47, such a Möbius transformation must be a similarity transformation, and so it is the composition of a central dilatation D and a linear isometry Ψ. Therefore, $T = \Phi D \Psi$, and each of the transformations on the right of this equation preserves the ratio ρ. The invariant ρ is the only one needed for Möbius classification, since any two profile spheres with the same value of ρ can be mapped to one another by a central dilatation. $\square$

Remark 5.20. We can obtain a family consisting of one representative from each Möbius equivalence class by fixing $a = 1$ and letting r vary, $0 < r < \infty$. This is just a family of parallel hypersurfaces of revolution. Taking a negative signed radius s for the profile sphere yields a parallel hypersurface that differs only in orientation from the hypersurface corresponding to $r = -s$. Finally, taking $r = 0$ also gives a parallel submanifold in the family, but the Euclidean projection degenerates to a sphere S^p. This is the case $\beta = 0, \alpha = e_{n+3}$, where the point sphere map equals the curvature sphere $[k_2]$ at every point.

5.5 Lie Frames

The goal of this section is to construct Lie frames that are well suited for the local study of Dupin submanifolds. These frames are fundamental in the local classification of Dupin hypersurfaces in $\mathbf{R}^4$ given in Section 5.7. They also provide a logical starting point for further local study of Dupin submanifolds, and indeed, this approach has been used by Cecil and Jensen [43], [44] and later by Cecil, Chi and Jensen [41] to study proper Dupin submanifolds with three or four distinct curvature spheres.

We will use the notation for the method of moving Lie frames introduced in Section 4.1. In particular, we agree on the following range of indices:

$$1 \le a, b, c \le n + 3, \qquad 3 \le i, j, k \le n + 1. \tag{5.56}$$

All summations are over the repeated index or indices. As in Section 4.1, a Lie frame is an ordered set of vectors $\{Y_1, \ldots, Y_{n+3}\}$ in $\mathbf{R}_2^{n+3}$ satisfying the relations,

$$\langle Y_a, Y_b \rangle = g_{ab} \tag{5.57}$$

for

$$[g_{ab}] = \begin{bmatrix} J & 0 & 0 \\ 0 & I_{n-1} & 0 \\ 0 & 0 & J \end{bmatrix}, \tag{5.58}$$

where I_{n-1} is the $(n-1) \times (n-1)$ identity matrix and

$$J = \begin{bmatrix} 0 & 1 \\ 1 & 0 \end{bmatrix}. \tag{5.59}$$

The Maurer–Cartan forms ω_a^b are defined by the equation

$$dY_a = \sum \omega_a^b Y_b. \tag{5.60}$$

Let $\lambda : M^{n-1} \to \Lambda^{2n-1}$ be an arbitrary Legendre submanifold. Let $\{Y_a\}$ be a smooth Lie frame on an open subset $U \subset M^{n-1}$ such that for each $x \in U$, we have

$$\lambda(x) = [Y_1(x), Y_{n+3}(x)].$$

We will pull back these Maurer–Cartan forms to U using the map λ^* and omit the symbols of such pull-backs for simplicity. Recall that the following matrix of forms is skew-symmetric,

$$[\omega_{ab}] = \begin{bmatrix} \omega_1^2 & \omega_1^1 & \omega_1^i & \omega_1^{n+3} & \omega_1^{n+2} \\ \omega_2^2 & \omega_2^1 & \omega_2^i & \omega_2^{n+3} & \omega_2^{n+2} \\ \omega_j^2 & \omega_j^1 & \omega_j^i & \omega_j^{n+3} & \omega_j^{n+2} \\ \omega_{n+2}^2 & \omega_{n+2}^1 & \omega_{n+2}^i & \omega_{n+2}^{n+3} & \omega_{n+2}^{n+2} \\ \omega_{n+3}^2 & \omega_{n+3}^1 & \omega_{n+3}^i & \omega_{n+3}^{n+3} & \omega_{n+3}^{n+2} \end{bmatrix}, \tag{5.61}$$

and that the forms satisfy the Maurer–Cartan equations,

$$d\omega_a^b = \sum \omega_a^c \wedge \omega_c^b. \tag{5.62}$$

By Theorem 4.8 of Chapter 4, p. 68, there are at most $n-1$ distinct curvature spheres along each line $\lambda(x)$. Thus, we can choose the Lie frame locally so that neither Y_1 nor Y_{n+3} is a curvature sphere at any point of U. Now $\omega_1^2 = 0$, by the skew-symmetry of the matrix in equation (5.61), and $\omega_1^{n+2} = 0$ by the Legendre condition (3) of Theorem 4.3 (p. 59) for λ. Thus, for any $X \in T_x M^{n-1}$ at any point $x \in U$, we have

$$dY_1(X) = \omega_1^1(X)Y_1 + \sum \omega_1^i(X)Y_i + \omega_1^{n+3}(X)Y_{n+3} \tag{5.63}$$

$$\equiv \sum \omega_1^i(X)Y_i, \quad \mod\{Y_1, Y_{n+3}\}.$$

The assumption that Y_1 is not a curvature sphere means that there does not exist any nonzero tangent vector X at any point $x \in U$ such that $dY_1(X)$ is congruent to zero, $\mod\{Y_1, Y_{n+3}\}$. By equation, (5.63), this assumption is equivalent to the condition that the forms $\{\omega_1^3, \dots, \omega_1^{n+1}\}$ be linearly independent, i.e., they satisfy the regularity condition

$$\omega_1^3 \wedge \cdots \wedge \omega_1^{n+1} \neq 0, \tag{5.64}$$

on U. Similarly, the condition that Y_{n+3} is not a curvature sphere is equivalent to the condition

$$\omega_{n+3}^3 \wedge \cdots \wedge \omega_{n+3}^{n+1} \neq 0. \tag{5.65}$$

We want to construct a Lie frame that is well-suited for the study of the curvature spheres of λ. The Legendre condition (3) for λ is $\omega_1^{n+2} = 0$. Exterior differentiation of this equation using equations (5.61)–(5.62) yields

$$\sum \omega_1^i \wedge \omega_{n+3}^i = 0. \tag{5.66}$$

Hence, by Cartan's lemma and the linear independence condition (5.64), we get that for each i,

$$\omega_{n+3}^i = \sum h_{ij}\omega_1^j, \quad \text{with } h_{ij} = h_{ji}. \tag{5.67}$$

The quadratic differential form

$$II(Y_1) = \sum h_{ij}\omega_1^i\omega_1^j, \tag{5.68}$$

defined up to a nonzero factor and dependent on the choice of Y_1, is called the *second fundamental form* of λ determined by Y_1.

To justify this name, we now consider the special case where Y_1 and Y_{n+3} are as follows:

$$Y_1 = (1 + f \cdot f, 1 - f \cdot f, 2f, 0)/2, \quad Y_{n+3} = (f \cdot \xi, -f \cdot \xi, \xi, 1), \quad (5.69)$$

where f is the Euclidean projection of λ, and ξ is the Euclidean field of unit normals. The condition (5.64) is equivalent to assuming that f is an immersion on U. Since f is an immersion, we can choose the Lie frame vectors $Y_3, \ldots, Y_{n+1}$ to satisfy

$$Y_i = dY_1(X_i) = (f \cdot df(X_i), -f \cdot df(X_i), df(X_i), 0), \quad 3 \le i \le n+1, \quad (5.70)$$

where $X_3, \ldots, X_{n+1}$ are smooth vector fields on U. Then we have

$$\omega_1^i(X_j) = \langle dY_1(X_j), Y_i \rangle = \langle Y_j, Y_i \rangle = \delta_{ij}. \tag{5.71}$$

Using equations (5.69) and (5.70), we compute

$$\omega_{n+3}^i(X_j) = \langle dY_{n+3}(X_j), Y_i \rangle = d\xi(X_j) \cdot df(X_i) \tag{5.72}$$
$$= -df(AX_j) \cdot df(X_i) = -A_{ij},$$

where $[A_{ij}]$ is the Euclidean shape operator (second fundamental form) of f. Now by equations (5.67) and (5.71), we have

$$\omega_{n+3}^i(X_j) = \sum h_{ik}\omega_1^k(X_j) = h_{ij}. \tag{5.73}$$

Thus, $h_{ij} = -A_{ij}$, whence the name "second fundamental form" for $[h_{ij}]$.

We now return to our general discussion, where λ is an arbitrary Legendre submanifold, and $\{Y_a\}$ is a Lie frame on U such that Y_1 and Y_{n+3} satisfy equations (5.64) and (5.65), respectively. Since $[h_{ij}]$ is symmetric, we can diagonalize it at any given point $x \in U$ by a change of frame of the form

$$Y_i^* = \sum C_i^j Y_j, \quad 3 \le i \le n+1,$$

where $[C_i^j]$ is an $(n-1) \times (n-1)$ orthogonal matrix. In the new frame, equation (5.67) has the following form at x,

$$\omega_{n+3}^i = -\mu_i \omega_1^i, \quad 3 \le i \le n+1. \tag{5.74}$$

These μ_i determine the curvature spheres of λ at x. Specifically, given any point $x \in U$, let

$$\{X_3, \ldots, X_{n+1}\}$$

be the dual basis to $\{\omega_1^3, \ldots, \omega_1^{n+1}\}$ in the tangent space $T_x M^{n-1}$. Then using equation (5.74), we compute the differential of $\mu_i Y_1 + Y_{n+3}$ on X_i to be

$$\begin{aligned}
d(\mu_i Y_1 + Y_{n+3})(X_i) &= d\mu_i(X_i)Y_1 + (\mu_i dY_1 + dY_{n+3})(X_i) \tag{5.75} \\
&\equiv \sum (\mu_i \omega_1^j(X_i) + \omega_{n+3}^j(X_i))Y_j \\
&= (\mu_i \omega_1^i(X_i) + \omega_{n+3}^i(X_i))Y_i \\
&= (\mu_i - \mu_i)Y_i = 0, \quad \mod\{Y_1, Y_{n+3}\}.
\end{aligned}$$

Hence, the curvature spheres of λ at x are precisely

$$K_i = \mu_i Y_1 + Y_{n+3}, \quad 3 \le i \le n+1, \tag{5.76}$$

and $X_3, \ldots, X_{n+1}$ are the principal vectors at x. In the case where Y_1 and Y_{n+3} have the form in equation (5.69), the μ_i are just the principal curvatures of the immersion f at the point x. From Corollary 4.9 of Chapter 4, p. 68, we know that if a curvature sphere K_i of the form (5.76) has constant multiplicity on U, then K_i and μ_i are both smooth on U, and the corresponding distribution of principal vectors is a foliation. As we noted after the proof of Corollary 4.9, there is an open dense subset of M^{n-1} on which the multiplicities of the curvature spheres are locally constant (see Reckziegel [157]–[158] or Singley [170]). For the remainder of this section, we assume that the number g of distinct curvature spheres is constant on U, and thus that each distinct curvature sphere has constant multiplicity on U. Assuming this, then the principal vector fields $X_3, \ldots, X_{n+1}$ can be chosen to be smooth on U. The frame vectors $Y_3, \ldots, Y_{n+1}$ can then be chosen to be smooth on U via the formula,

$$Y_i = dY_1(X_i), \quad 3 \le i \le n+1.$$

As in equation (5.71), this means that $\{\omega_1^3, \ldots, \omega_1^{n+1}\}$ is the dual basis to $\{X_3, \ldots, X_{n+1}\}$. Equation (5.74) is then satisfied at every point of U. This frame is an example of what we will call a principal frame. In general, a Lie frame $\{Z_a\}$ on U is said to be a *principal Lie frame* if there exist smooth functions α_i and β_i on U, which are never simultaneously zero, such that the Maurer–Cartan forms $\{\theta_a^b\}$ for the frame satisfy the equations,

$$\alpha_i \theta_1^i + \beta_i \theta_{n+3}^i = 0, \quad 3 \le i \le n+1. \tag{5.77}$$

It is worth noting that θ_1^i and θ_{n+3}^i cannot both vanish at a point x in U. To see this, take a Lie frame $\{W_a\}$ on U with

$$W_i = Z_i, \quad 3 \le i \le n+1,$$

such that $W_1 = \alpha Z_1 + \beta Z_{n+3}$ is not a curvature sphere at x. Then the Maurer–Cartan form ϕ_1^i for this frame satisfies the equation

$$\phi_1^i = \langle dW_1, W_i \rangle = \langle \alpha Z_1 + \beta Z_{n+3}, Z_i \rangle = \alpha \theta_1^i + \beta \theta_{n+3}^i.$$

Since W_1 is not a curvature sphere, it follows that $\phi_1^i \neq 0$, and thus it is not possible for θ_1^i and θ_{n+3}^i to both equal zero.

There is a fair amount of flexibility in the choice of principal Lie frame. We next show how the choice can be made more specific in order to have the Maurer–Cartan forms give more direct information about a given curvature sphere. We can assume that $\{Y_a\}$ is a principal frame on U satisfying equations (5.64) and (5.65) and that the curvature spheres are given by equation (5.76). In particular, suppose that

$$K = \mu Y_1 + Y_{n+3}$$

is a curvature sphere of multiplicity m on U. Then the function μ is smooth on U, and we can reorder the frame vectors $Y_3, \ldots, Y_{n+1}$ so that

$$\mu = \mu_3 = \cdots = \mu_{m+2} \tag{5.78}$$

on U. Since Y_1 and Y_{n+3} are not curvature spheres at any point of U, the function μ never takes the value 0 or ∞ on U. We now make a change of frame so that $Y_1^* = K$ is a curvature sphere of multiplicity m. Specifically, let

$$Y_1^* = \mu Y_1 + Y_{n+3}, \qquad\qquad Y_2^* = (1/\mu)Y_2, \tag{5.79}$$
$$Y_{n+2}^* = Y_{n+2} - (1/\mu)Y_2, \qquad Y_{n+3}^* = Y_{n+3}, \qquad Y_i^* = Y_i, \quad 3 \leq i \leq n+1.$$

Denote the Maurer–Cartan forms for this frame by θ_a^b. Note that

$$dY_1^* = d(\mu Y_1 + Y_{n+3}) = (d\mu)Y_1 + \mu \, dY_1 + dY_{n+3} = \sum \theta_1^a Y_a^*. \tag{5.80}$$

Using equation (5.74), we see that the coefficient of $Y_i^* = Y_i$ in equation (5.80) is

$$\theta_1^i = \mu \omega_1^i + \omega_{n+3}^i = (\mu - \mu_i)\omega_1^i, \quad 3 \leq i \leq n+1. \tag{5.81}$$

This and equation (5.78) show that

$$\theta_1^r = 0, \quad 3 \leq r \leq m+2. \tag{5.82}$$

Equation (5.82) characterizes the condition that Y_1^* is a curvature sphere of constant multiplicity m on U.

We now want to study the Dupin condition that $K = Y_1^*$ is constant along each leaf of its corresponding principal foliation. As noted in Corollary 4.9 of Chapter 4, p. 68, this is automatic if the multiplicity m of K is greater than one. We denote this principal foliation by T_1 and choose smooth vector fields

$$\{X_3, \ldots, X_{m+2}\}$$

on U that span T_1. The condition that Y_1^* be constant along each leaf of its principal foliation is

$$dY_1^*(X_r) \equiv 0, \quad \mathrm{mod}\, Y_1^*, \quad 3 \le r \le m+2. \tag{5.83}$$

On the other hand, from equations (5.71) and (5.81), we have

$$dY_1^*(X_r) = \theta_1^1(X_r)Y_1 + \theta_1^{n+3}(X_r)Y_{n+3}, \quad 3 \le r \le m+2. \tag{5.84}$$

Comparing equations (5.83) and (5.84), we see that

$$\theta_1^{n+3}(X_r) = 0, \quad 3 \le r \le m+2. \tag{5.85}$$

We now show that we can make one more change of frame so that in the new frame the Maurer–Cartan form $\alpha_1^{n+3} = 0$. We can write θ_1^{n+3} in terms of the basis $\{\omega_1^3, \dots, \omega_1^{n+1}\}$ as

$$\theta_1^{n+3} = \sum s_i \omega_1^i, \tag{5.86}$$

for smooth functions s_i on U. From equation (5.85), we see that we actually have

$$\theta_1^{n+3} = \sum_{t=m+3}^{n+1} s_t \omega_1^t. \tag{5.87}$$

Using equations (5.71), (5.81) and (5.87), we compute that for $m+3 \le t \le n+1$,

$$\begin{aligned} dY_1^*(X_t) &= \theta_1^1(X_t)Y_1^* + \theta_1^t(X_t)Y_t + \theta_1^{n+3}(X_t)Y_{n+3} \\ &= \theta_1^1(X_t)Y_1^* + (\mu - \mu_t)Y_t + s_t Y_{n+3} \\ &= \theta_1^1(X_t)Y_1^* + (\mu - \mu_t)(Y_t + (s_t/(\mu - \mu_t))Y_{n+3}). \end{aligned} \tag{5.88}$$

We now make the change of Lie frame,

$$\begin{aligned} Z_1 &= Y_1^*, \quad Z_2 = Y_2^*, \quad Z_{n+3} = Y_{n+3}^*, \quad Z_r = Y_r^* = Y_r, \quad 3 \le r \le m+2, \\ Z_t &= Y_t + (s_t/(\mu - \mu_t))Y_{n+3}, \quad m+3 \le t \le n+1, \\ Z_{n+2} &= -\sum_t (s_t/(\mu - \mu_t))Y_t + Y_{n+2} - (1/2)\sum_t (s_t/(\mu - \mu_t))^2 Y_{n+3}. \end{aligned} \tag{5.89}$$

Let α_a^b be the Maurer–Cartan forms for this new frame. We still have

$$\alpha_1^r = \langle dZ_1, Z_r \rangle = \langle dY_1^*, Y_r^* \rangle = \theta_1^r = 0, \quad 3 \le r \le m+2. \tag{5.90}$$

Furthermore, since $Z_1 = Y_1^*$, the Dupin condition (5.83) still yields

$$\alpha_1^{n+3}(X_r) = 0, \quad 3 \le r \le m+2. \tag{5.91}$$

Finally, for $m+3 \le t \le n+1$, equations (5.88) and (5.89) yield

$$\alpha_1^{n+3}(X_t) = \langle dZ_1(X_t), Z_{n+2} \rangle = \langle \theta_1^1(X_t)Z_1 + (\mu - \mu_t)Z_t, Z_{n+2} \rangle = 0. \tag{5.92}$$

Thus, we have $\alpha_1^{n+3} = 0$. We summarize these results in the following theorem.

Theorem 5.21. *Let* $\lambda : M^{n-1} \to \Lambda^{2n-1}$ *be a Legendre submanifold. Suppose that K is a curvature sphere of multiplicity m on an open subset U of M^{n-1} that is constant along each leaf of its principal foliation. Then locally on U, there exists a Lie frame $\{Y_1, \ldots, Y_{n+3}\}$ with $Y_1 = K$, such that the Maurer–Cartan forms satisfy the equations*

$$\omega_1^r = 0, \quad 3 \leq r \leq m+2, \quad \omega_1^{n+3} = 0. \tag{5.93}$$

5.6 Covariant Differentiation

In this section we discuss a method of covariant differentiation, classically known as "half-invariant" differentiation, which is sometimes useful in simplifying the computation of derivatives in the projective context. This method is discussed in complete generality in the book of Bol [11, Vol. III, pp. 1–15]. Here we show how it can be used in the study of Legendre submanifolds. The method of half-invariant differentiation plays an important role in Pinkall's [149] local classification of Dupin hypersurfaces in $\mathbf{R}^4$ given in the next section, and in later works by Niebergall [126]–[127], and Cecil and Chern [38], Cecil and Jensen [44]–[45], and Cecil, Chi, and Jensen [41].

Let $\lambda : M^{n-1} \to \Lambda^{2n-1}$ be a Legendre submanifold and suppose that $\{Y_a\}$ is a Lie frame defined on an open subset U of M^{n-1}. One degree of freedom that is always present in the choice of a Lie frame is a *renormalization* of the form

$$\begin{aligned}
Y_1^* &= \tau Y_1, & Y_2^* &= (1/\tau)Y_2, \\
Y_{n+3}^* &= \tau Y_{n+3}, & Y_{n+2}^* &= (1/\tau)Y_{n+2}, \\
Y_i^* &= Y_i, & 3 &\leq i \leq n+1,
\end{aligned} \tag{5.94}$$

where τ is a smooth nonvanishing function on U. This is equivalent to a "star-transformation," as defined by Pinkall [146], [149]. If Y_1 is not a curvature sphere at any point of U, then the forms

$$\{\omega_1^3, \ldots, \omega_1^{n+1}\}$$

constitute a basis for the cotangent space T^*U at each point of U. For a principal frame, it may be that some of the ω_1^i are zero. In that case, as we showed in the previous section, the corresponding form ω_{n+3}^i is not zero, and one can still choose a basis locally for T^*U so that each element of the basis is either a ω_1^i or a ω_{n+3}^i. We assume that we have such a basis, which we denote by $\{\theta_1, \ldots, \theta_{n-1}\}$. Under a renormalization of the form (5.94), we have for $3 \leq i \leq n+1$,

$$\begin{aligned}
\omega_1^{i*} &= \langle dY_1^*, Y_i^* \rangle = \langle d(\tau Y_1), Y_i \rangle = \langle (d\tau)Y_1 + \tau \, dY_1, Y_i \rangle \\
&= \tau \langle dY_1, Y_i \rangle = \tau \omega_1^i.
\end{aligned} \tag{5.95}$$

In a similar way, we have $\omega_{n+3}^{i*} = \tau \omega_{n+3}^i$, for $3 \leq i \leq n+1$. Thus, all of the elements of the basis transform by the formula

$$\theta_r^* = \tau \theta_r, \quad 1 \leq r \leq n-1. \tag{5.96}$$

The form ω_1^1 transforms as follows under renormalization,

$$
\begin{aligned}
\omega_1^{1*} &= \langle dY_1^*, Y_2^* \rangle = \langle d(\tau Y_1), (1/\tau)Y_2 \rangle = (1/\tau)\langle \tau \, dY_1 + (d\tau)Y_1, Y_2 \rangle \\
&= \langle dY_1, Y_2 \rangle + (d\tau/\tau)\langle Y_1, Y_2 \rangle = \omega_1^1 + d(\log \tau).
\end{aligned}
\tag{5.97}
$$

Hence, the form $\pi = -\omega_1^1$ transforms as a *Reeb form* for the renormalization [11], i.e.,

$$
\pi^* = \pi - d(\log \tau).
\tag{5.98}
$$

Such a Reeb form is crucial in the covariant (or half-invariant) differentiation process defined below.

A scalar function or differential form ω on U is said to be *half-invariant of weight* m if there is an integer m such that under renormalization, ω transforms by the formula

$$
\omega^* = \tau^m \omega.
\tag{5.99}
$$

Thus the basis forms θ_r, $1 \le r \le n - 1$, are all half-invariant of weight one. For two half-invariant forms, ω_1 and ω_2, we set

$$
(\omega_1 \wedge \omega_2)^* = \omega_1^* \wedge \omega_2^*.
$$

If ω_1 has weight m_1, and ω_2 has weight m_2, then by equation (5.99) the product $\omega_1 \wedge \omega_2$ has weight $m_1 + m_2$. This is useful in determining the weight of certain functions which arise naturally. For example, if $3 \le i \le n + 1$, then the form ω_i^1 has weight -1, since

$$
\omega_i^{1*} = \langle dY_i^*, Y_2^* \rangle = \langle dY_i, (1/\tau)Y_2 \rangle = (1/\tau)\langle dY_i, Y_2 \rangle = (1/\tau)\omega_i^1.
$$

Now suppose that we write ω_i^1 in terms of the basis $\{\theta_r\}$ as

$$
\omega_i^1 = a_1 \theta_1 + \cdots + a_{n-1}\theta_{n-1}.
$$

Then each of the functions a_r has weight -2, since the weight of a_r plus the weight of θ_r must equal -1.

The covariant derivative or *half-invariant derivative* $\tilde{d}\omega$ of a function or differential form ω of weight m is defined by the formula

$$
\tilde{d}\omega = d\omega + m\,\pi \wedge \omega,
\tag{5.100}
$$

where π is the Reeb form. We see that $\tilde{d}\omega$ is also half-invariant of weight m as follows. From equation (5.99), we compute

$$
d\omega^* = \tau^m d\omega + m\,\tau^m d(\log \tau) \wedge \omega.
\tag{5.101}
$$

Then using equation (5.98), we have

$$
\tilde{d}\omega^* = d\omega^* + m\,\pi^* \wedge \omega^* = \tau^m(d\omega + m\,\pi \wedge \omega) = \tau^m \tilde{d}\omega.
\tag{5.102}
$$

Using the rules for ordinary exterior differentiation, it is straightforward to verify the following rules for this covariant differentiation process. Suppose that ω_1 and ω_2 are half-invariant forms and that ω_1 is a q-form, then

$$\tilde{d}(\omega_1 \wedge \omega_2) = \tilde{d}\omega_1 \wedge \omega_2 + (-1)^q \omega_1 \wedge \tilde{d}\omega_2. \tag{5.103}$$

If ω_1 and ω_2 are of the same weight, then

$$\tilde{d}(\omega_1 + \omega_2) = \tilde{d}\omega_1 + \tilde{d}\omega_2. \tag{5.104}$$

Unlike ordinary exterior differentiation, differentiating twice by this method does not lead to zero, in general. In fact, if ω has weight m, then

$$\tilde{d}(\tilde{d}\omega) = d(d\omega + m\,\pi \wedge \omega) + m\,\pi \wedge (d\omega + m\,\pi \wedge \omega). \tag{5.105}$$

Using equation (5.103) and the equation $d(d\omega) = 0$, one can reduce the equation above to

$$\tilde{d}(\tilde{d}\omega) = m\,d\pi \wedge \omega. \tag{5.106}$$

From equation (5.106), we see that $\tilde{d}(\tilde{d}\omega) = 0$ for every form ω only when $d\pi = 0$. In that case, since we are working locally, we can take U to be contractible, and thus $\pi = d\sigma$, for some smooth scalar function σ on U. If we take a renormalization with $\tau = e^\sigma$, then we have from equation (5.98) that $\pi^* = 0$. Equation (5.100) implies that after this renormalization, half-invariant differentiation is just ordinary exterior differentiation. This special choice of τ is determined up to a constant factor. In cases where $d\pi \neq 0$, there may not be such an optimal choice of renormalization. In those cases, $\tilde{d}\tilde{d} \neq 0$, but there are still certain commutation relations which hold for the computing of second derivatives. These are of primary importance.

For the rest of this section, we adopt the following convention on the indices:

$$1 \leq i, j, k \leq n - 1. \tag{5.107}$$

Suppose now that f is a half-invariant scalar function of weight m on U. Using the fact that $\tilde{d}f$ has weight m and that the basis forms have weight one, we can write

$$\tilde{d}f = f_1\theta_1 + \cdots + f_{n-1}\theta_{n-1}, \tag{5.108}$$

where the f_i are half-invariant functions of weight $m - 1$. We call the f_i the *covariant derivatives* or *half-invariant derivatives of f with respect to the basis* $\{\theta_i\}$. We now derive the commutation relations which are satisfied by the covariant derivatives of these f_i. Using equation (5.106), we take the covariant derivative of equation (5.108) and get

$$mf\,d\pi = \sum \tilde{d}f_i \wedge \theta_i + f_i\,\tilde{d}\theta_i. \tag{5.109}$$

We express the 2-forms $d\pi$ and $\tilde{d}\theta_i$ in terms of the basis forms as follows:

$$d\pi = \sum_{i<j} p_{ij}\,\theta_i \wedge \theta_j, \qquad \tilde{d}\theta_k = \sum_{i<j} c_{ij}^k\,\theta_i \wedge \theta_j. \tag{5.110}$$

For convenience in subsequent formulas, we also define the coefficients for the case $i \geq j$ by setting

$$p_{ji} = -p_{ij}, \qquad c_{ji}^k = -c_{ij}^k \quad \text{for all } i, j, k. \tag{5.111}$$

By equation (5.108), we have

$$\tilde{d} f_i = \sum_j f_{ij}\, \theta_j. \tag{5.112}$$

If we now substitute equation (5.110) and (5.112) into equation (5.109), we obtain the following equation:

$$mf \sum_{i<j} p_{ij}\theta_i \wedge \theta_j = \sum_i \sum_j f_{ij}\, \theta_j \wedge \theta_i + \sum_k f_k\, \tilde{d}\theta_k \tag{5.113}$$

$$= \sum_i \sum_j f_{ij}\, \theta_j \wedge \theta_i + \sum_k f_k \sum_{i<j} c_{ij}^k\, \theta_i \wedge \theta_j.$$

By taking the coefficient of $\theta_i \wedge \theta_j$ in equation (5.113), we obtain the *commutation relations*

$$f_{ij} - f_{ji} = \sum_k f_k c_{ij}^k - mf p_{ij}. \tag{5.114}$$

By the convention (5.111), these relations are valid for all i, j. Thus, the commutation relations for the second derivatives of a half-invariant function f can be determined from the p_{ij}, c_{ij}^k, f and the first derivatives of f. This is crucial in the solution of the differential equations involved in a typical application of this method. The classification of Dupin hypersurfaces in $\mathbf{R}^4$ in the next section is a good example.

5.7 Dupin Hypersurfaces in 4-Space

In this section, we give Pinkall's [146], [149] local classification of proper Dupin hypersurfaces in $\mathbf{R}^4$ up to Lie equivalence (see also Cecil–Chern [38]). The umbilic case with $g = 1$ distinct curvature sphere is well known, and the hypersurface must be an open subset of a hypersphere or a hyperplane in $\mathbf{R}^4$. The case $g = 2$ is handled by Theorem 5.18. Thus, the only case remaining is $g = 3$. The classification in that case is rather involved, and it makes using of the method of moving frames in a way that was not necessary in the case $g = 2$. It is the first case where Lie invariants are necessary in the classification, and it is worthy of careful study. Later classification results of Niebergall [126], [127], Cecil and Jensen [44], [45], and Cecil, Chi and Jensen [41] use a similar approach to that employed here.

We will follow the notation of the previous section, and consider a proper Dupin submanifold $\lambda : M^3 \to \Lambda^7$ with three distinct curvature spheres (of multiplicity one) at each point. Locally, λ is Lie equivalent to a Dupin hypersurface immersed in $\mathbf{R}^4$, but we do not assume that the Euclidean projection of λ into $\mathbf{R}^4$ is an immersion. As before, we make a local choice of Lie frame $\{Y_1, \ldots, Y_7\}$ on an open subset U of M^3 so that for each $x \in M^3$, we have

$$\lambda(x) = [Y_1(x), Y_7(x)]. \qquad (5.115)$$

We can take Y_1 and Y_7 to be curvature spheres at each point of U. Furthermore, by applying Theorem 5.21 first to Y_1 and then to Y_7, we can arrange that the Maurer–Cartan forms for the Lie frame satisfy

$$\omega_1^3 = 0, \qquad \omega_1^7 = 0, \qquad \omega_7^4 = 0, \qquad \omega_7^1 = 0. \qquad (5.116)$$

Next by making a change of the form

$$Y_1^* = \sigma Y_1, \qquad Y_2^* = (1/\sigma)Y_2, \qquad Y_7^* = \tau Y_7, \qquad Y_6^* = (1/\tau)Y_6, \qquad (5.117)$$

for suitable smooth functions σ and τ on U, we can arrange that $Y_1 + Y_7$ represents the third curvature sphere at each point of U. Then we can use the method of proof of Theorem 5.21 to find a Lie frame whose Maurer–Cartan forms satisfy

$$\omega_1^5 + \omega_7^5 = 0, \qquad \omega_1^1 - \omega_7^7 = 0, \qquad (5.118)$$

as well as equation (5.116). Such a frame is called a *second order frame* in the terminology of Cecil and Jensen [44, p. 138]. Conditions (5.116) and (5.118) completely determine Y_3, Y_4, and Y_5, while Y_1 and Y_7 are determined up to a transformation of the form

$$Y_1^* = \tau Y_1, \qquad Y_7^* = \tau Y_7, \qquad (5.119)$$

for some smooth nonvanishing function τ on U, i.e., a renormalization.

Each of the three curvature sphere maps Y_1, Y_7 and $Y_1 + Y_7$ is constant along each leaf of its corresponding principal foliation. Thus, each of these maps factors through an immersion of the corresponding 2-dimensional space of leaves of its principal foliation into the Lie quadric Q^5. In terms of moving frames, this implies that the forms ω_1^4, ω_1^5 and ω_7^3 are linearly independent, i.e.,

$$\omega_1^4 \wedge \omega_1^5 \wedge \omega_7^3 \neq 0. \qquad (5.120)$$

This can also be seen by expressing the forms above in terms of a Lie frame $\{Z_1, \ldots, Z_{n+3}\}$ whose Maurer–Cartan forms satisfy the regularity condition (5.64). For simplicity, we will also use the notation as in Section 5.6,

$$\theta_1 = \omega_1^4, \qquad \theta_2 = \omega_1^5, \qquad \theta_3 = \omega_7^3. \qquad (5.121)$$

Analytically, the Dupin conditions are three partial differential equations, and we are treating an overdetermined system. The method of moving frames reduces the handling of its integrability conditions to a straightforward algebraic problem, viz., that of repeated exterior differentiations. Later, we will use the method of half-invariant differentiation to simplify some of the calculations.

We begin by computing the exterior derivatives of the equations,

$$\omega_1^3 = 0, \qquad \omega_7^4 = 0, \qquad \omega_1^5 + \omega_7^5 = 0. \tag{5.122}$$

These equations come from the fact that Y_1, Y_7 and $Y_1 + Y_7$ are curvature spheres. Using the skew-symmetry of the matrix in equation (5.61), as well as the relations (5.116) and (5.118), the exterior derivatives of the three equations in (5.122) yield

$$0 = \omega_1^4 \wedge \omega_3^4 + \omega_1^5 \wedge \omega_3^5,$$
$$0 = \omega_1^5 \wedge \omega_4^5 + \omega_7^3 \wedge \omega_3^4, \tag{5.123}$$
$$0 = \omega_1^4 \wedge \omega_4^5 + \omega_7^3 \wedge \omega_3^5.$$

If we take the wedge product of the first of these equations with ω_1^4, we conclude that ω_3^5 is in the span of ω_1^4 and ω_1^5. On the other hand, taking the wedge product of the third equation with ω_1^4 yields that ω_3^5 is in the span of ω_1^4 and ω_7^3. Consequently, $\omega_3^5 = \rho \omega_1^4$, for some smooth function ρ. Similarly, there exist smooth functions α and β such that $\omega_3^4 = \alpha \omega_1^5$ and $\omega_4^5 = \beta \omega_7^3$. Then, if we substitute these results into equation (5.123), we get that $\rho = \alpha = \beta$, and hence we have

$$\omega_3^5 = \rho \omega_1^4, \qquad \omega_3^4 = \rho \omega_1^5, \qquad \omega_4^5 = \rho \omega_7^3. \tag{5.124}$$

This function ρ plays a crucial role in the classification. Next we differentiate the three equations that come from the Dupin conditions,

$$\omega_1^7 = 0, \qquad \omega_7^1 = 0, \qquad \omega_1^1 - \omega_7^7 = 0. \tag{5.125}$$

As above, the use of the skew-symmetry relations and equations (5.116) and (5.118) yields the existence of smooth functions $a, b, c, p, q, r, s, t, u$ such that the following relations hold:

$$\omega_4^7 = -\omega_6^4 = a\omega_1^4 + b\omega_1^5, \tag{5.126}$$
$$\omega_5^7 = -\omega_6^5 = b\omega_1^4 + c\omega_1^5;$$
$$\omega_3^1 = -\omega_2^3 = p\omega_7^3 - q\omega_1^5, \tag{5.127}$$
$$\omega_5^1 = -\omega_2^5 = q\omega_7^3 - r\omega_1^5;$$
$$\omega_4^1 = -\omega_2^4 = b\omega_1^5 + s\omega_1^4 + t\omega_7^3, \tag{5.128}$$
$$\omega_6^3 = -\omega_3^7 = q\omega_1^5 + t\omega_1^4 + u\omega_7^3.$$

We next take the exterior derivatives of the three basis forms ω_1^4, ω_1^5 and ω_7^3. Using the relations that we have derived so far, we obtain from the Maurer–Cartan equation (5.62),

$$d\omega_1^4 = \omega_1^1 \wedge \omega_1^4 + \omega_1^5 \wedge \omega_5^4 = \omega_1^1 \wedge \omega_1^4 - \rho \omega_1^5 \wedge \omega_7^3. \tag{5.129}$$

We obtain similar expressions for $d\omega_1^5$ and $d\omega_7^3$ in the same way. Now when we write these expressions in terms of θ_1, θ_2 and θ_3, they become

$$d\theta_1 = \omega_1^1 \wedge \theta_1 - \rho\, \theta_2 \wedge \theta_3,$$

$$d\theta_2 = \omega_1^1 \wedge \theta_2 - \rho \, \theta_3 \wedge \theta_1, \tag{5.130}$$
$$d\theta_3 = \omega_1^1 \wedge \theta_3 - \rho \, \theta_1 \wedge \theta_2.$$

As we noted in Section 5.6, the form $\pi = -\omega_1^1$ can always be used as a Reeb form for half-invariant differentiation. Using the formula,

$$\tilde{d}\omega = d\omega + m \, \pi \wedge \omega,$$

for the half-invariant derivative of a form ω of weight m, and recalling that the θ_i have weight one, we can rewrite equation (5.130) as

$$\tilde{d}\theta_1 = -\rho \, \theta_2 \wedge \theta_3,$$
$$\tilde{d}\theta_2 = -\rho \, \theta_3 \wedge \theta_1, \tag{5.131}$$
$$\tilde{d}\theta_3 = -\rho \, \theta_1 \wedge \theta_2.$$

From equations (5.110), (5.111) and (5.131), we see that the functions c_{ij}^k for the half-invariant differentiation have the form

$$c_{23}^1 = c_{31}^2 = c_{12}^3 = -\rho, \quad c_{32}^1 = c_{13}^2 = c_{21}^3 = \rho, \quad c_{ij}^k = 0 \quad \text{otherwise.} \tag{5.132}$$

We next differentiate equation (5.124). We have $\omega_3^4 = \rho \omega_1^5$. On the one hand,

$$d\omega_3^4 = \rho \, d\omega_1^5 + d\rho \wedge \omega_1^5.$$

Using the second equation in (5.130) with $\omega_1^5 = \theta_2$, this becomes

$$d\omega_3^4 = \rho \omega_1^1 \wedge \omega_1^5 - \rho^2 \omega_7^3 \wedge \omega_1^4 + d\rho \wedge \omega_1^5.$$

On the other hand, we can compute $d\omega_3^4$ from the Maurer–Cartan equation (5.62) and use the relationships that we have derived to find

$$d\omega_3^4 = (-p - \rho^2 - a)(\omega_1^4 \wedge \omega_7^3) - q\omega_1^5 \wedge \omega_1^4 + b\omega_7^3 \wedge \omega_1^5.$$

Equating these two expressions for $d\omega_3^4$ yields

$$(-p - a - 2\rho^2)(\omega_1^4 \wedge \omega_7^3) = (d\rho + \rho\omega_1^1 - q\omega_1^4 - b\omega_7^3) \wedge \omega_1^5. \tag{5.133}$$

Due to the linear independence of the forms $\{\omega_1^4, \omega_1^5, \omega_7^3\}$, both sides of the equation above must vanish. Thus, we conclude that

$$2\rho^2 = -a - p, \tag{5.134}$$

and that $d\rho + \rho\omega_1^1 - q\omega_1^4 - b\omega_7^3$ is a multiple of ω_1^5. Similarly, differentiation of the equation $\omega_4^5 = \rho\omega_7^3$ yields the following analogue of equation (5.133),

$$(s - a - r + 2\rho^2)\omega_1^4 \wedge \omega_1^5 = (d\rho + \rho\omega_1^1 + t\omega_1^5 - q\omega_1^4) \wedge \omega_7^3, \tag{5.135}$$

and differentiation of $\omega_3^5 = \rho\omega_1^4$ yields

$$(c + p + u - 2\rho^2)\omega_1^5 \wedge \omega_7^3 = (-d\rho - \rho\omega_1^1 - t\omega_1^5 + b\omega_7^3) \wedge \omega_1^4. \qquad (5.136)$$

In each of the equations (5.133), (5.135), (5.136), both sides of the equation must vanish. From the vanishing of the left sides of the equations, we get the fundamental relationship

$$2\rho^2 = -a - p = a + r - s = c + p + u. \qquad (5.137)$$

Furthermore, from the vanishing of the right sides of the three equations (5.133), (5.135), (5.136), we can determine after some algebra that

$$d\rho + \rho\omega_1^1 = q\omega_1^4 - t\omega_1^5 + b\omega_7^3. \qquad (5.138)$$

This equation shows the importance of ρ as follows. From equation (5.124), we see that the half-invariant function ρ has weight -1, since ω_3^5 has weight zero and ω_1^4 has weight one. Recalling that $\pi = -\omega_1^1$, we have

$$d\rho + \rho\omega_1^1 = d\rho - \rho\pi = \tilde{d}\rho = \rho_1\theta_1 + \rho_2\theta_2 + \rho_3\theta_3. \qquad (5.139)$$

Comparing equations (5.138) and (5.139), we find that the half-invariant derivatives ρ_i are given by

$$\rho_1 = q, \quad \rho_2 = -t, \quad \rho_3 = b. \qquad (5.140)$$

Using the Maurer–Cartan equation (5.62), we can compute

$$\begin{aligned} d\omega_1^1 &= \omega_1^4 \wedge \omega_4^1 + \omega_1^5 \wedge \omega_5^1 \qquad &(5.141)\\ &= \omega_1^4 \wedge (b\omega_1^5 + t\omega_7^3) + \omega_1^5 \wedge (q\omega_7^3 - r\omega_1^5) \\ &= b\omega_1^4 \wedge \omega_1^5 + q\omega_1^5 \wedge \omega_7^3 - t\omega_7^3 \wedge \omega_1^4. \end{aligned}$$

Using equations (5.121) and (5.140), this can be rewritten as

$$d\omega_1^1 = \rho_3\,\theta_1 \wedge \theta_2 + \rho_1\,\theta_2 \wedge \theta_3 + \rho_2\,\theta_3 \wedge \theta_1. \qquad (5.142)$$

Since $\pi = -\omega_1^1$, comparing equations (5.142) and (5.110) yields the following formulas for the coefficients p_{ij} of the half-invariant derivative of π,

$$p_{21} = -p_{12} = \rho_3, \quad p_{32} = -p_{23} = \rho_1, \quad p_{13} = -p_{31} = \rho_2. \qquad (5.143)$$

By equations (5.132) and (5.143), the fundamental coefficients for the half-invariant differentiation process are all expressed in terms of ρ and its half-invariant derivatives ρ_i. This enables us to express the half-invariant derivatives of all the functions involved in terms of ρ and its successive half-invariant derivatives. Ultimately, this leads to the solution of the problem.

First note that equations (5.132) and (5.143) allow us to write the commutation relations (5.114) for a half-invariant function σ of weight m in the form

$$\sigma_{12} - \sigma_{21} = -\rho\sigma_3 + m\sigma\rho_3,$$

$$\sigma_{23} - \sigma_{32} = -\rho\sigma_1 + m\sigma\rho_1, \tag{5.144}$$
$$\sigma_{31} - \sigma_{13} = -\rho\sigma_2 + m\sigma\rho_2.$$

In particular, since ρ is half-invariant of weight -1, we have

$$\rho_{12} - \rho_{21} = -2\rho\rho_3$$
$$\rho_{23} - \rho_{32} = -2\rho\rho_1 \tag{5.145}$$
$$\rho_{31} - \rho_{13} = -2\rho\rho_2.$$

We next take the exterior derivative of equations (5.126)–(5.128). We first differentiate the equation

$$\omega_4^7 = a\omega_1^4 + b\omega_1^5. \tag{5.146}$$

On the one hand, from the Maurer–Cartan equation (5.62) for $d\omega_4^7$, and omitting those terms that are already known to vanish, we have

$$d\omega_4^7 = \omega_4^2 \wedge \omega_2^7 + \omega_4^3 \wedge \omega_3^7 + \omega_4^5 \wedge \omega_5^7 + \omega_4^7 \wedge \omega_7^7 \tag{5.147}$$
$$= -\omega_1^4 \wedge \omega_2^7 + (-\rho)\omega_1^5 \wedge (-q\omega_1^5 - t\omega_1^4 - u\omega_7^3)$$
$$+ \rho\omega_7^3 \wedge (b\omega_1^4 + c\omega_1^5) + (a\omega_1^4 + b\omega_1^5) \wedge \omega_1^1.$$

On the other hand, differentiation of the right side of equation (5.146) yields

$$d\omega_4^7 = da \wedge \omega_1^4 + ad\omega_1^4 + db \wedge \omega_1^5 + bd\omega_1^5 \tag{5.148}$$
$$= da \wedge \omega_1^4 + a(\omega_1^1 \wedge \omega_1^4 - \rho\omega_1^5 \wedge \omega_7^3)$$
$$+ db \wedge \omega_1^5 + b(\omega_1^1 \wedge \omega_1^5 - \rho\omega_1^4 \wedge \omega_7^3).$$

Equating (5.147) and (5.148), we find

$$(da + 2a\omega_1^1 - 2b\rho\omega_7^3 - \omega_2^7) \wedge \omega_1^4 \tag{5.149}$$
$$+ (db + 2b\omega_1^1 + (a + u - c)\rho\omega_7^3) \wedge \omega_1^5 + \rho t\omega_1^4 \wedge \omega_1^5 = 0.$$

Since $b = \rho_3$ is half-invariant of weight -2, we have

$$db + 2b\omega_1^1 = d\rho_3 + 2\rho_3\omega_1^1 = \tilde{d}\rho_3 = \rho_{31}\theta_1 + \rho_{32}\theta_2 + \rho_{33}\theta_3. \tag{5.150}$$

By examining the coefficient of $\omega_1^5 \wedge \omega_7^3 = \theta_2 \wedge \theta_3$ in equation (5.149) and using equation (5.150), we find

$$\rho_{33} = \rho(c - a - u). \tag{5.151}$$

Furthermore, the remaining terms in equation (5.149) are

$$(da + 2a\omega_1^1 - \omega_2^7 - 2\rho b\omega_7^3 - (\rho t + \rho_{31})\omega_1^5) \wedge \omega_1^4 \tag{5.152}$$
$$+ \text{ terms involving } \omega_1^5 \text{ and } \omega_7^3 \text{ only.}$$

Thus, the coefficient in parentheses must be a multiple of ω_1^4, call it $\bar{a}\omega_1^4$. Since

$$\omega_4^7 = \langle dY_4, Y_6 \rangle$$

has weight -1, and ω_1^4 has weight one, it follows from equation (5.126) that the function a has weight -2. Using equation (5.140), we can write equation (5.152) as

$$da + 2a\omega_1^1 = da - 2a\pi = \tilde{d}a = \omega_2^7 + \bar{a}\theta_1 + (\rho_{31} - \rho\rho_2)\theta_2 + 2\rho\rho_3\theta_3. \quad (5.153)$$

In a similar manner, if we differentiate the equation

$$\omega_5^7 = b\omega_1^4 + c\omega_1^5,$$

and use the fact that the function c has weight -2, we obtain

$$dc + 2c\omega_1^1 = dc - 2c\pi = \tilde{d}c = \omega_2^7 + (\rho_{32} + \rho\rho_1)\theta_1 + \bar{c}\theta_2 - 2\rho\rho_3\theta_3. \quad (5.154)$$

Thus, from the two equations in (5.126), we have obtained equations (5.151), (5.153) and (5.154). In a completely analogous manner, we can differentiate the two equations in (5.127) to obtain

$$\rho_{11} = \rho(s + r - p), \quad (5.155)$$

$$dp + 2p\omega_1^1 = \tilde{d}p = -\omega_2^7 + 2\rho\rho_1\theta_1 + (-\rho_{13} - \rho\rho_2)\theta_2 + \bar{p}\theta_3, \quad (5.156)$$

$$dr + 2r\omega_1^1 = \tilde{d}r = -\omega_2^7 - 2\rho\rho_1\theta_1 + \bar{r}\theta_2 + (-\rho_{12} + \rho\rho_3)\theta_3. \quad (5.157)$$

Similarly, differentiation of equation (5.128) yields

$$\rho_{22} + \rho_{33} = \rho(p - r - s), \quad (5.158)$$

$$ds + 2s\omega_1^1 = \tilde{d}s = \bar{s}\theta_1 + (\rho_{31} + \rho\rho_2)\theta_2 + (-\rho_{21} + \rho\rho_3)\theta_3, \quad (5.159)$$

$$du + 2u\omega_1^1 = \tilde{d}u = (-\rho_{23} - \rho\rho_1)\theta_1 + (\rho_{13} - \rho\rho_2)\theta_2 + \bar{u}\theta_3. \quad (5.160)$$

In these equations, the coefficients $\bar{a}, \bar{c}, \bar{p}, \bar{r}, \bar{s}$ and $\bar{u}$ remain undetermined. However, by differentiating equation (5.137) and using the appropriate equations from above, one can show that

$$\begin{aligned} \bar{a} &= -6\rho\rho_1, & \bar{c} &= 6\rho\rho_2, \\ \bar{p} &= -6\rho\rho_3, & \bar{r} &= 6\rho\rho_2, \\ \bar{s} &= -12\rho\rho_1, & \bar{u} &= 12\rho\rho_3. \end{aligned} \quad (5.161)$$

From equations (5.151), (5.155), (5.158) and (5.137), we can easily compute that

$$\rho_{11} + \rho_{22} + \rho_{33} = 0. \quad (5.162)$$

Using equation (5.161), we can rewrite equations (5.159) and (5.160) as

$$ds + 2s\omega_1^1 = \tilde{d}s = -12\rho\rho_1\theta_1 + (\rho_{31} + \rho\rho_2)\theta_2 + (-\rho_{21} + \rho\rho_3)\theta_3, \quad (5.163)$$

$$du + 2u\omega_1^1 = \tilde{d}u = (-\rho_{23} - \rho\rho_1)\theta_1 + (\rho_{13} - \rho\rho_2)\theta_2 + 12\rho\rho_3\theta_3. \quad (5.164)$$

Thus the half-invariant derivatives of s and u are expressed in terms of ρ and its half-invariant derivatives. By taking half-invariant derivatives of these two equations and making use of equation (5.162) and the commutation relations (5.144) for ρ and its various derivatives, one can show after a lengthy calculation that the following fundamental equations hold:

$$\rho\rho_{12} + \rho_1\rho_2 + \rho^2\rho_3 = 0,$$
$$\rho\rho_{21} + \rho_1\rho_2 - \rho^2\rho_3 = 0,$$
$$\rho\rho_{23} + \rho_2\rho_3 + \rho^2\rho_1 = 0,$$
$$\rho\rho_{32} + \rho_2\rho_3 - \rho^2\rho_1 = 0, \tag{5.165}$$
$$\rho\rho_{31} + \rho_3\rho_1 + \rho^2\rho_2 = 0,$$
$$\rho\rho_{13} + \rho_3\rho_1 - \rho^2\rho_2 = 0.$$

We now briefly outline the details of this calculation. By equation (5.163), we have

$$s_1 = -12\rho\rho_1, \qquad s_2 = \rho_{31} + \rho\rho_2, \qquad s_3 = \rho\rho_3 - \rho_{21}. \tag{5.166}$$

The half-invariant quantity s has weight -2, so the commutation relation (5.144) gives

$$s_{12} - s_{21} = -2s\rho_3 - \rho s_3 = -2s\rho_3 - \rho(\rho\rho_3 - \rho_{21}). \tag{5.167}$$

On the other hand, by taking the half-invariant derivatives of equation (5.166), we can compute directly that

$$s_{12} - s_{21} = -12\rho\rho_{12} - 12\rho_2\rho_1 - (\rho_{311} + \rho_1\rho_2 + \rho\rho_{21}). \tag{5.168}$$

The main problem now is to get the half-invariant derivative ρ_{311} into a workable form. By taking the half-invariant derivative of the third equation in (5.145), we find

$$\rho_{311} - \rho_{131} = -2\rho_1\rho_2 - 2\rho\rho_{21}. \tag{5.169}$$

Then using the commutation relation,

$$\rho_{131} = \rho_{113} - 2\rho_1\rho_2 - \rho\rho_{12},$$

we get from equation (5.169) that

$$\rho_{311} = \rho_{113} - 4\rho_1\rho_2 - \rho\rho_{12} - 2\rho\rho_{21}. \tag{5.170}$$

Taking the half-invariant derivative of the equation,

$$\rho_{11} = \rho(s + r - p),$$

and substituting the expression obtained for ρ_{113} into equation (5.170), we get

$$\rho_{311} = \rho_3(s + r - p) - 3\rho\rho_{21} - 2\rho\rho_{12} + 8\rho^2\rho_3 - 4\rho_1\rho_2. \tag{5.171}$$

If we substitute this expression for ρ_{311} into equation (5.168) and then equate the right sides of equations (5.167) and (5.168), we obtain the first equation in (5.165). The cyclic permutations are obtained in a similar way from $s_{23} - s_{32}$, and so on.

Although Y_3, Y_4 and Y_5 are already completely determined by the conditions (5.116) and (5.118), it is still possible to make a change of frame of the form

$$Y_1^* = \tau Y_1, \qquad Y_2^* = (1/\tau)Y_2 + \mu Y_7, \qquad (5.172)$$
$$Y_7^* = \tau Y_7, \qquad Y_6^* = (1/\tau)Y_6 - \mu Y_1.$$

Under this change of frame, we have

$$\omega_1^{4*} = \tau \omega_1^4, \qquad \omega_1^{5*} = \tau \omega_1^5, \qquad \omega_7^{3*} = \tau \omega_7^3,$$
$$\omega_4^{7*} = (1/\tau)\omega_4^7 + \mu \omega_1^4, \qquad (5.173)$$
$$\omega_3^{1*} = (1/\tau)\omega_3^1 - \mu \omega_7^3.$$

Suppose that we write

$$\omega_4^{7*} = a^* \omega_1^{4*} + b^* \omega_1^{5*}, \qquad (5.174)$$
$$\omega_3^{1*} = p^* \omega_7^{3*} - q^* \omega_1^{5*}.$$

Then from equation (5.173), we obtain

$$a^* = \tau^{-2}a + \tau^{-1}\mu, \qquad p^* = \tau^{-2}p - \tau^{-1}\mu.$$

Thus, by taking $\mu = (p - a)/2\tau$, we can arrange that $a^* = p^*$. We now make this change of frame and drop the asterisks. In this new frame, we have

$$a = p = -\rho^2, \qquad r = 3\rho^2 + s, \qquad c = 3\rho^2 - \mu. \qquad (5.175)$$

Using the fact that $a = p$, we can subtract equation (5.156) from equation (5.153) and get that

$$\omega_2^7 = 4\rho\rho_1\theta_1 - ((\rho_{31} + \rho_{13})/2)\theta_2 - 4\rho\rho_3\theta_3. \qquad (5.176)$$

Now through equations (5.153)–(5.157), the half-invariant derivatives of the functions a, c, p and r are expressed in terms of ρ and its derivatives. We are now ready to proceed to the main results. Ultimately, we show that it is possible to choose a frame in which ρ is constant. Thus, the classification naturally splits into two cases, $\rho = 0$ and $\rho \neq 0$. We handle the two cases separately.

CASE 1. $\rho \neq 0$ (the irreducible case).

Assume that the function ρ is never zero on the open set U on which the frame $\{Y_a\}$ is defined. The key step in getting ρ to be constant is the following lemma due to Pinkall [149, p. 108], where his function c is the negative of our function ρ. The formulation of the proof here was first given in Cecil–Chern [38, p. 33], and it is somewhat simpler than Pinkall's proof. The crucial point here is that since $\rho \neq 0$, the fundamental equations (5.165) allow us to express all of the second half-invariant derivatives ρ_{ij} of ρ in terms of ρ and its first derivatives.

Lemma 5.22. *Suppose that the function ρ never vanishes on the open set U on which the frame $\{Y_a\}$ is defined. Then its half-invariant derivatives satisfy $\rho_1 = \rho_2 = \rho_3 = 0$ at every point of U.*

Proof. First, note that if ρ_3 vanishes identically, then the equations (5.165) and the assumption that $\rho \neq 0$ imply that ρ_1 and ρ_2 also vanish identically. We now complete the proof by showing that ρ_3 must vanish everywhere on U. This is accomplished by considering the expression $s_{12} - s_{21}$. By the commutation relations (5.144), we have

$$s_{12} - s_{21} = -2s\rho_3 - \rho s_3.$$

By equations (5.165)–(5.166), we see that

$$\rho s_3 = \rho^2 \rho_3 - \rho \rho_{21} = \rho_1 \rho_2,$$

and so

$$s_{12} - s_{21} = -2s\rho_3 - \rho_1 \rho_2. \qquad (5.177)$$

On the other hand, we can compute s_{12} by taking the derivative of the equation

$$s_1 = -12\rho\rho_1.$$

Then using the expression for ρ_{12} obtained from equation (5.165), we get

$$s_{12} = -12\rho_2 \rho_1 - 12\rho\rho_{12} = -12(\rho_2 \rho_1 + \rho \rho_{12}) \qquad (5.178)$$
$$= -12(\rho_2 \rho_1 + (-\rho_2 \rho_1 - \rho^2 \rho_3)) = 12\rho^2 \rho_3.$$

Next we have from equation (5.166) that $s_2 = \rho_{31} + \rho\rho_2$. Using equation (5.165), we have

$$\rho_{31} = -\rho_3 \rho_1 \rho^{-1} - \rho\rho_2,$$

and thus,

$$s_2 = -\rho_3 \rho_1 / \rho. \qquad (5.179)$$

Then we compute

$$s_{21} = -(\rho(\rho_3 \rho_{11} + \rho_{31} \rho_1) - \rho_3 \rho_1^2)/\rho^2.$$

Using equation (5.155) for ρ_{11} and (5.165) to get ρ_{31}, this becomes

$$s_{21} = -\rho_3(s + r - p) + 2\rho_3 \rho_1^2 \rho^{-2} + \rho_1 \rho_2. \qquad (5.180)$$

Now equate the expression in equation (5.177) for $s_{12} - s_{21}$ with the expression obtained by subtracting equation (5.180) from equation (5.178) to get

$$-2s\rho_3 - \rho_1 \rho_2 = 12\rho^2 \rho_3 + \rho_3(s + r - p) - 2\rho_3 \rho_1^2 \rho^{-2} - \rho_1 \rho_2.$$

This can be rewritten as

$$0 = \rho_3(12\rho^2 + 3s + r - p - 2\rho_1^2 \rho^{-2}). \qquad (5.181)$$

Using the expressions in equations (5.175) for r and p, we see that

$$3s + r - p = 4s + 4\rho^2,$$

and so, equation (5.181) can be rewritten as

$$0 = \rho_3(16\rho^2 + 4s - 2\rho_1^2\rho^{-2}). \tag{5.182}$$

Suppose that $\rho_3 \neq 0$ at some point x of U. Then ρ_3 does not vanish on some neighborhood V of x. By equation (5.182), we have

$$16\rho^2 + 4s - 2\rho_1^2\rho^{-2} = 0 \tag{5.183}$$

on V. We now take the θ_2-half-invariant derivative of equation (5.183) and obtain

$$32\rho\rho_2 + 4s_2 - 4\rho_1\rho_{12}\rho^{-2} + 4\rho_1^2\rho_2\rho^{-3} = 0. \tag{5.184}$$

We can now substitute the expression (5.179) for s_2 and the formula

$$\rho_{12} = -\rho_1\rho_2\rho^{-1} - \rho\rho_3$$

obtained from equation (5.165) into equation (5.184). After some algebra, equation (5.184) reduces to

$$\rho_2(32\rho^4 + 8\rho_1^2) = 0.$$

Since $\rho \neq 0$, this implies that $\rho_2 = 0$ on V. But then, the left side of the equation below, obtained from (5.165),

$$\rho\rho_{21} + \rho_1\rho_2 = \rho^2\rho_3,$$

must vanish on V. Since $\rho \neq 0$, we conclude that $\rho_3 = 0$ on V, a contradiction to our assumption. Hence, ρ_3 must vanish identically on the set U, and the lemma is proved.   □

We now continue with the case $\rho \neq 0$. According to the previous lemma, all the half-invariant derivatives of ρ are zero, and our formulas simplify greatly. Equations (5.151) and (5.155) give

$$c - a - u = 0, \qquad s + r - p = 0.$$

These combined with equation (5.175) give

$$c = r = \rho^2, \qquad u = -s = 2\rho^2. \tag{5.185}$$

By equation (5.176), we have $\omega_2^7 = 0$. So the differentials of the frame vectors can now be written as

$$dY_1 - \omega_1^1 Y_1 = \omega_1^4 Y_4 + \omega_1^5 Y_5,$$
$$dY_7 - \omega_1^1 Y_7 = \omega_7^3 Y_3 - \omega_1^5 Y_5,$$

$$dY_2 + \omega_1^1 Y_2 = \rho^2(\omega_7^3 Y_3 + 2\omega_1^4 Y_4 + \omega_1^5 Y_5),$$
$$dY_6 + \omega_1^1 Y_6 = \rho^2(2\omega_7^3 Y_3 + \omega_1^4 Y_4 - \omega_1^5 Y_5),$$
$$dY_3 = \omega_7^3 Z_3 + \rho(\omega_1^5 Y_4 + \omega_1^4 Y_5),$$
$$dY_4 = -\omega_1^4 Z_4 + \rho(-\omega_1^5 Y_3 + \omega_7^3 Y_5),$$
$$dY_5 = \omega_1^5 Z_5 + \rho(-\omega_1^4 Y_3 - \omega_7^3 Y_4),$$

(5.186)

where

$$Z_3 = -Y_6 + \rho^2(-Y_1 - 2Y_7),$$
$$Z_4 = Y_2 + \rho^2(2Y_1 + Y_7),$$
$$Z_5 = -Y_2 + Y_6 + \rho^2(-Y_1 + Y_7).$$

(5.187)

From this we notice that

$$Z_3 + Z_4 + Z_5 = 0,$$

(5.188)

so that the points Z_3, Z_4 and Z_5 lie on a line in projective space $\mathbf{P}^6$. From equations (5.139), (5.142) and the lemma above, we see that

$$\tilde{d}\rho = d\rho + \rho\omega_1^1 = d\rho - \rho\pi = 0, \quad d\omega_1^1 = -d\pi = 0.$$

(5.189)

Here we have the special situation $d\pi = 0$, which we discussed after equation (5.106) in Section 5.6. In that case, there is always a renormalization (5.94) determined up to a constant factor, which makes $\pi^* = 0$. From equation (5.189), we have

$$\pi = d(\log \rho).$$

Hence, we take the renormalization factor

$$\tau = e^{\log \rho} = \rho.$$

This makes $\pi^* = \omega_1^{1*} = 0$. Furthermore, since the function ρ, as defined in equation (5.124), has weight -1, we also get that

$$\rho^* = \rho^{-1}\rho = 1$$

in the new frame. Thus, we now make a change of frame of the form

$$Y_1^* = \rho Y_1, \qquad Y_2^* = (1/\rho)Y_2,$$
$$Y_7^* = \rho Y_7, \qquad Y_6^* = (1/\rho)Y_6,$$
$$Y_i^* = Y_i, \qquad\quad 3 \le i \le 5.$$

(5.190)

We see from equation (5.187) that the Z_i are half-invariant of weight -1. Further, the basis forms ω_1^4, ω_1^5 and ω_7^3 are of weight one. Thus, we have

$$Z_i^* = (1/\rho)Z_i, \quad 3 \le i \le 5,$$

(5.191)

$$\omega_1^{4*} = \rho\omega_1^4, \qquad \omega_1^{5*} = \rho\omega_1^5, \qquad \omega_7^{3*} = \rho\omega_7^3.$$

Using the equations above, we compute the differentials of the frame vectors as follows:

$$
\begin{aligned}
dY_1^* &= \omega_1^{4*}Y_4 + \omega_1^{5*}Y_5, \\
dY_7^* &= \omega_7^{3*}Y_3 - \omega_1^{5*}Y_5, \\
dY_2^* &= \omega_7^{3*}Y_3 + 2\omega_1^{4*}Y_4 + \omega_1^{5*}Y_5, \\
dY_6^* &= 2\omega_7^{3*}Y_3 + \omega_1^{4*}Y_4 - \omega_1^{5*}Y_5, \\
dY_3 &= \omega_7^{3*}Z_3^* + \omega_1^{5*}Y_4 + \omega_1^{4*}Y_5, \\
dY_4 &= -\omega_1^{4*}Z_4^* - \omega_1^{5*}Y_3 + \omega_7^{3*}Y_5, \\
dY_5 &= \omega_1^{5*}Z_5^* - \omega_1^{4*}Y_3 - \omega_7^{3*}Y_4,
\end{aligned}
\tag{5.192}
$$

with

$$
\begin{aligned}
dZ_3^* &= 2(-2\omega_7^{3*}Y_3 - \omega_1^{4*}Y_4 + \omega_1^{5*}Y_5), \\
dZ_4^* &= 2(\omega_7^{3*}Y_3 + 2\omega_1^{4*}Y_4 + \omega_1^{5*}Y_5), \\
dZ_5^* &= 2(\omega_7^{3*}Y_3 - \omega_1^{4*}Y_4 - 2\omega_1^{5*}Y_5),
\end{aligned}
\tag{5.193}
$$

and

$$
\begin{aligned}
d\omega_1^{4*} &= -\omega_1^{5*} \wedge \omega_7^{3*}, \quad \text{i.e.,} \quad d\theta_1^* = -\theta_2^* \wedge \theta_3^*, \\
d\omega_1^{5*} &= -\omega_7^{3*} \wedge \omega_1^{4*}, \quad \text{i.e.,} \quad d\theta_2^* = -\theta_3^* \wedge \theta_1^*, \\
d\omega_7^{3*} &= -\omega_1^{4*} \wedge \omega_1^{5*}, \quad \text{i.e.,} \quad d\theta_3^* = -\theta_1^* \wedge \theta_2^*.
\end{aligned}
\tag{5.194}
$$

Comparing the last equation with equation (5.130), we see that

$$\omega_1^{1*} = 0, \qquad \rho^* = 1. \tag{5.195}$$

This is the final frame needed in the case $\rho \neq 0$, so we drop the asterisks once more.

We can now prove Pinkall's classification for the case $\rho \neq 0$. As with the cyclides of Dupin, there is only one model up to Lie equivalence. This model is Cartan's isoparametric hypersurface with three principal curvatures in S^4. Cartan's hypersurface is a tube over each of its two focal submanifolds in S^4, both of which are Veronese surfaces. (See [52, pp. 296–299] for more detail.) After we prove the following theorem, we will show that Cartan's classification of isoparametric hypersurfaces with three principal curvatures in S^4 can be derived using our methods.

Theorem 5.23.

(a) *Every connected Dupin proper submanifold*

$$\lambda : M^3 \to \Lambda^7$$

with three distinct curvature spheres and $\rho \neq 0$ is contained in a unique compact, connected proper Dupin submanifold with $\rho \neq 0$.

(b) *Any two proper Dupin submanifolds with $\rho \neq 0$ are locally Lie equivalent, each being Lie equivalent to an open subset of an isoparametric hypersurface in S^4.*

Proof. Let $\{Y_a\}$ be the Lie frame just constructed on a connected open subset $U \subset M^3$ satisfying (5.195), i.e.,

$$\omega_1^1 = 0, \qquad \rho = 1. \tag{5.196}$$

Then the derivatives of the frame vectors satisfy the system of equations (5.192), where we again drop the asterisks. The three curvature sphere maps on U are Y_1, Y_7 and $Y_1 + Y_7$. Let

$$W_1 = -Y_1 + Y_6 - 2Y_7, \qquad W_2 = -2Y_1 + Y_2 - Y_7. \tag{5.197}$$

Then from equation (5.192), we find that

$$dW_1 = dW_2 = 0.$$

Hence W_1 and W_2 are constant maps. Furthermore, since

$$\langle W_1, W_1 \rangle = \langle W_2, W_2 \rangle = -4, \qquad \langle W_1, W_2 \rangle = -2,$$

the line $[W_1, W_2]$ is timelike. Finally, the equations,

$$\langle Y_1, W_1 \rangle = 0, \qquad \langle Y_7, W_2 \rangle = 0, \qquad \langle Y_1 + Y_7, W_1 - W_2 \rangle = 0, \tag{5.198}$$

imply that the restriction of λ to U is Lie equivalent to an open subset of an isoparametric hypersurface in S^4 by Theorem 4.19 of Chapter 4, p. 77.

If $\{\tilde{Y}_a\}$ is a Lie frame defined on an open subset $\tilde{U} \subset M^3$ by the same construction as $\{Y_a\}$, and $U \cap \tilde{U}$ is nonempty, then by the uniqueness of the construction, at points of $U \cap \tilde{U}$ the curvature spheres satisfy

$$\tilde{Y}_1 = Y_1, \qquad \tilde{Y}_7 = Y_7, \qquad \tilde{Y}_1 + \tilde{Y}_7 = Y_1 + Y_7,$$

and the points $\tilde{W}_1 = W_1$, $\tilde{W}_2 = W_2$. Thus, the timelike line $[W_1, W_2]$ and the points W_1 and W_2 on it satisfying equation (5.198) are the same on the set $\tilde{U}$ as they are on U, and hence they are the same on all of the connected manifold M^3. Therefore, the whole Dupin submanifold $\lambda : M^3 \to \Lambda^7$ is Lie equivalent to an open subset of an isoparametric hypersurface in S^4. Since any connected open subset of an isoparametric hypersurface is contained in a unique compact, connected isoparametric hypersurface (see Münzner [123] or Cecil–Ryan [52, Sections 4–6 of Chapter 3]), part (a) is proved. Furthermore, because all isoparametric hypersurfaces in S^4 are locally Lie equivalent by a result of Cartan [17], part (b) is also true. □

Theorem 5.23 relies on Cartan's classification of isoparametric hypersurfaces in S^4 for the completion of its proof. We will now show directly using our methods that a connected Dupin proper submanifold $\lambda : U^3 \to \Lambda^7$ with three distinct curvature spheres and $\rho \neq 0$ is Lie equivalent to an open subset of the Legendre submanifold induced by a Veronese surface. Together with the proof of Theorem 5.23, this gives a proof of Cartan's classification of isoparametric hypersurfaces in S^4. The proof of Theorem 5.24 below was first given in the paper of Cecil–Chern [38].

For our purposes here, we use the following definition of a Veronese surface. This definition is slightly different than the one given in Example 5.6, but it is equivalent. Let S^2 be the unit sphere in $\mathbf{R}^3$ given by the equation

$$y_1^2 + y_2^2 + y_3^2 = 1.$$

Consider the map from S^2 into the unit sphere $\mathbf{R}^5$ given by

$$(y_1, y_2, y_3) \mapsto (2y_2y_3, 2y_3y_1, 2y_1y_2, y_1^2, y_2^2). \tag{5.199}$$

This map takes the same value on antipodal points of S^2, so it induces a map ϕ : $\mathbf{P}^2 \to \mathbf{R}^5$. One can show that ϕ is an embedding of $\mathbf{P}^2$ and that ϕ is substantial in $\mathbf{R}^5$, i.e., the image of ϕ does not lie in any hyperplane in $\mathbf{R}^5$. Of course, ϕ can also be considered as an embedding of $\mathbf{P}^2$ into $\mathbf{P}^5$ by considering $\mathbf{R}^5$ as an affine space in $\mathbf{P}^5$. Any embedding of $\mathbf{P}^2$ into $\mathbf{P}^5$ which is projectively equivalent to ϕ is called a Veronese surface. (See Lane [100, pp. 424–430] for more detail.) A Veronese surface is said to be *spherical* if its image lies in the unit sphere $S^4 \subset \mathbf{R}^5$.

Theorem 5.24. *Every connected Dupin proper submanifold* $\lambda : M^3 \to \Lambda^7$ *with three distinct curvature spheres and* $\rho \neq 0$ *is Lie equivalent to an open subset of the Legendre submanifold induced by a spherical Veronese surface*

$$V \subset S^4 \subset \mathbf{R}^5.$$

Proof. We continue with the same notation that we have been using in this section. Thus, $\{Y_a\}$ is the final Lie frame defined on a connected open subset $U \subset M^3$, and Y_1, Y_7 and $Y_1 + Y_7$ are the three curvature sphere maps of λ. Each curvature sphere map is constant along the leaves of its principal foliation, so each curvature sphere map factors through an immersion of the two-dimensional space of leaves of its principal foliation. We will show that each of these immersions is an open subset of a Veronese surface in some $\mathbf{P}^5 \subset \mathbf{P}^6$.

We wish to integrate the system of differential equations (5.192), which is completely integrable. So we drop the asterisks, and using (5.121) rewrite the system as

$$
\begin{aligned}
dY_1 &= \theta_1 Y_4 + \theta_2 Y_5, \\
dY_7 &= \theta_3 Y_3 - \theta_2 Y_5, \\
dY_2 &= \theta_3 Y_3 + 2\theta_1 Y_4 + \theta_2 Y_5, \\
dY_6 &= 2\theta_3 Y_3 + \theta_1 Y_4 - \theta_2 Y_5, \\
dY_3 &= \theta_3 Z_3 + \theta_2 Y_4 + \theta_1 Y_5, \\
dY_4 &= -\theta_1 Z_4 - \theta_2 Y_3 + \theta_3 Y_5, \\
dY_5 &= \theta_2 Z_5 - \theta_1 Y_3 - \theta_3 Y_4,
\end{aligned}
\tag{5.200}
$$

where equation (5.187) yields the following values for Z_3, Z_4, Z_5, since $\rho = 1$,

$$Z_3 = -Y_1 - Y_6 - 2Y_7,$$
$$Z_4 = 2Y_1 + Y_2 + Y_7, \tag{5.201}$$
$$Z_5 = -Y_1 - Y_2 + Y_6 + Y_7.$$

As before, we have

$$Z_3 + Z_4 + Z_5 = 0, \tag{5.202}$$

Equations (5.193) and (5.194) become

$$dZ_3 = 2(-2\theta_3 Y_3 - \theta_1 Y_4 + \theta_2 Y_5),$$
$$dZ_4 = 2(\theta_3 Y_3 + 2\theta_1 Y_4 + \theta_2 Y_5), \tag{5.203}$$
$$dZ_5 = 2(\theta_3 Y_3 - \theta_1 Y_4 - 2\theta_2 Y_5),$$

and

$$d\theta_1 = -\theta_2 \wedge \theta_3,$$
$$d\theta_2 = -\theta_3 \wedge \theta_1, \tag{5.204}$$
$$d\theta_3 = -\theta_1 \wedge \theta_2.$$

As in equation (5.197), we let

$$W_1 = -Y_1 + Y_6 - 2Y_7, \qquad W_2 = -2Y_1 + Y_2 - Y_7. \tag{5.205}$$

Then as we showed above, $dW_1 = dW_2 = 0$, so that the maps W_1 and W_2 are both constant, and the line $[W_1, W_2]$ determined by the points W_1 and W_2 consists entirely of timelike points. The orthogonal complement of this line in $\mathbf{R}_2^7$ is the 5-dimensional spacelike vector space,

$$\mathbf{R}^5 = \text{Span}\{Y_3, Y_4, Y_5, Z_4, Z_5\}.$$

It suffices to solve the system in equation (5.200) in $\mathbf{R}^5$ for Y_3, Y_4, Y_5, Z_4, Z_5. This is true because we have

$$d(Z_4 - Z_5 - 6Y_1) = 0, \qquad d(Z_4 + 2Z_5 - 6Y_7) = 0, \tag{5.206}$$

so that there exist constant vectors C_1 and C_2 such that

$$Z_4 - Z_5 - 6Y_1 = C_1, \qquad Z_4 + 2Z_5 - 6Y_7 = C_2. \tag{5.207}$$

Thus, Y_1 and Y_7 are determined by these equations, and then Y_2 and Y_6 are determined by equation (5.205). Note that C_1 and C_2 are timelike points, and the line $[C_1, C_2]$ consists entirely of timelike points.

The equations (5.204) are the structure equations of the group $SO(3)$. Thus, it is natural to take $SO(3)$ as the parameter space. The group $SO(3)$ consists of 3×3 real matrices

$$A = [a_{ik}], \quad 1 \le i, j, k \le 3,$$

satisfying the equations

$$A^t A = A A^t = I, \quad \det A = 1. \tag{5.208}$$

The first equations above, when written in terms of the entries of the matrix A, are

$$\sum a_{ij} a_{ik} = \sum a_{ji} a_{ki} = \delta_{jk}, \tag{5.209}$$

where we again employ the convention that all summations are over the repeated index or indices. The Maurer–Cartan forms of $SO(3)$ are

$$\alpha_{ik} = \sum a_{kj} da_{ij} = -\alpha_{ki}. \tag{5.210}$$

These forms satisfy the Maurer–Cartan equations,

$$d\alpha_{ik} = \sum \alpha_{ij} \wedge \alpha_{jk}. \tag{5.211}$$

If we set

$$\theta_1 = \alpha_{23}, \qquad \theta_2 = \alpha_{31}, \qquad \theta_3 = \alpha_{12}, \tag{5.212}$$

then these Maurer–Cartan equations reduce to (5.204). With the θ_i as given in equation (5.212), we can now find an explicit solution to the system (5.200) as follows.

Let $\{E_1, \ldots, E_5\}$ be a fixed linear frame in $\mathbf{R}^5$. We define

$$F_i = 2a_{i2}a_{i3}E_1 + 2a_{i3}a_{i1}E_2 + 2a_{i1}a_{i2}E_3 + a_{i1}^2 E_4 + a_{i2}^2 E_5, \tag{5.213}$$

for $1 \leq i, j, k \leq 3$. Since

$$a_{i1}^2 + a_{i2}^2 + a_{i3}^2 = 1,$$

we see from equation (5.199), that F_i is a Veronese surface for $1 \leq i, j, k \leq 3$. Using equation (5.209), we compute that

$$F_1 + F_2 + F_3 = E_4 + E_5 = \text{constant}. \tag{5.214}$$

Since the coefficients in F_i are quadratic, the second partial derivatives,

$$\frac{\partial^2 F_i}{\partial a_{ij} \partial a_{ik}},$$

are independent of i. Moreover, the quantities

$$G_{ik} = \sum_{j=1}^{3} a_{ij} \frac{\partial F_k}{\partial a_{kj}} \tag{5.215}$$

satisfy the relation

$$G_{ik} = G_{ki}. \tag{5.216}$$

We make use of these facts in the following computation:

$$dG_{ik} = \sum \frac{\partial F_k}{\partial a_{kj}} da_{ij} + \sum a_{ij} \frac{\partial^2 F_k}{\partial a_{kj} \partial a_{kl}} da_{kl} \tag{5.217}$$

$$= \sum \frac{\partial F_k}{\partial a_{kj}} da_{ij} + \sum a_{ij} \frac{\partial^2 F_i}{\partial a_{ij} \partial a_{il}} da_{kl}$$

$$= \sum \frac{\partial F_k}{\partial a_{kj}} da_{ij} + \sum \frac{\partial F_i}{\partial a_{ij}} da_{kj},$$

where the last step follows from the linear homogeneity of $\partial F_i / \partial a_{il}$. In terms of α_{ij}, we have

$$dG_{ik} = \sum \frac{\partial F_k}{\partial a_{kj}} a_{lj} \alpha_{il} + \sum \frac{\partial F_i}{\partial a_{ij}} a_{lj} \alpha_{kl}, \tag{5.218}$$

which gives, when expanded,

$$dG_{23} = 2(F_3 - F_1)\theta_1 + G_{12}\theta_2 - G_{13}\theta_3, \tag{5.219}$$

and its cyclic permutations.

On the other hand, we have by the same manipulation,

$$dF_i = \sum \frac{\partial F_i}{\partial a_{ij}} da_{ij} = \sum \frac{\partial F_i}{\partial a_{ij}} a_{kj} \alpha_{ik}, \tag{5.220}$$

which gives

$$dF_1 = -G_{31}\theta_2 + G_{12}\theta_3,$$

and its cyclic permutations.

One can now immediately verify that a solution to the system (5.200) is given by

$$
\begin{array}{lll}
Y_3 = G_{12}, & Y_4 = -G_{23}, & Y_5 = G_{31}, \\
Z_3 = 2(F_2 - F_1), & Z_4 = 2(F_3 - F_2), & Z_5 = 2(F_1 - F_3).
\end{array} \tag{5.221}
$$

This is also the most general solution of the system (5.200), for the solution is determined up to a linear transformation, and our choice of frame $\{E_1, \ldots, E_5\}$ is arbitrary.

By equation (5.221), the functions Z_1, Z_2, and Z_3 are expressible in terms of F_1, F_2 and F_3, and thus by equation (5.207), so also are Y_1, Y_7 and $Y_1 + Y_7$. Specifically, by equations (5.207), (5.214) and (5.221), we have

$$6Y_1 = Z_4 - Z_5 - C_1 = 2(-F_1 - F_2 + 2F_3) - C_1 \tag{5.222}$$

$$= 6F_3 - 2(E_4 + E_5) - C_1,$$

so that the curvature sphere map Y_1, up to an additive constant, is the Veronese surface F_3. Similarly, the curvature sphere maps Y_7 and $Y_1 + Y_7$ are the Veronese surfaces F_1 and $-F_2$, respectively, up to additive constants.

From equation (5.205), we see that

$$\langle Y_1, W_1 \rangle = 0, \qquad \langle Y_7, W_2 \rangle = 0, \qquad \langle Y_1 + Y_7, W_1 - W_2 \rangle = 0. \tag{5.223}$$

Thus, Y_1 is contained in the Möbius space $\Sigma^4 = Q^5 \cap W_1^\perp$. Let

$$e_1 = (2W_2 - W_1)/\sqrt{12}, \qquad e_7 = W_1/2.$$

Then e_1 is the unit vector on the timelike line $[W_1, W_2]$ that is orthogonal to W_1. In a manner similar to Section 4.2, we can write

$$Y_1 = e_1 + f, \tag{5.224}$$

where f maps U into the unit sphere S^4 in the Euclidean space

$$\mathbf{R}^5 = [e_1, e_7]^{\perp} \subset \mathbf{R}_2^7.$$

Thus, f is the spherical projection of the Legendre map λ onto this sphere S^4. We know that f is constant along each leaf of the principal foliation T_1 corresponding to the curvature sphere Y_1, and that f induces an immersion on the space of leaves U/T_1,

$$\tilde{f} : U/T_1 \rightarrow S^4.$$

By what we have shown above, $\tilde{f}$ is an open subset of a spherical Veronese surface.

Note that the unit timelike vector $W_2/2$ satisfies the equation

$$W_2/2 = (\sqrt{3}/2)e_1 + (1/2)e_7 = \sin(\pi/3)e_1 + \cos(\pi/3)e_7.$$

If we consider the points in the Möbius space Σ to represent point spheres in S^4, then it follows from equation (2.21) of Chapter 2, p. 17, that the points in $Q^5 \cap W_2^{\perp}$ represent oriented spheres in S^4 of signed radius $-\pi/3$. In a manner similar to that above, we can show that the curvature sphere map Y_7 induces a spherical Veronese surface that lies in the space $Q^5 \cap W_2^{\perp}$. When considered from the point of view of the Möbius space Σ, the points in the image of Y_7 represent oriented spheres of signed radius $-\pi/3$ centered at points of this Veronese surface. These spheres are in oriented contact with the point spheres of the first Veronese surface determined by Y_1 in $Q^5 \cap \mathbf{P}^5$. Thus the points in the second Veronese surface must lie at a distance $\pi/3$ along normal geodesics in S^4 from the first Veronese surface $\tilde{f}$. In fact (see, for example [52, pp. 296–299]), the set of all points in S^4 at a distance $\pi/3$ from a spherical Veronese surface is another spherical Veronese surface.

Thus, with this choice of coordinates, the Dupin submanifold λ is simply an open subset of the Legendre submanifold induced from the Veronese surface $\tilde{f}$ as a submanifold of codimension 2 in S^4. For values of $t = k\pi/3, k \in \mathbf{Z}$, the parallel submanifold at oriented distance t from $\tilde{f}$ is a Veronese surface. For other values of t, the parallel submanifold is the Legendre submanifold induced by an isoparametric hypersurface in S^4 with three principal curvatures (Cartan's isoparametric hypersurface). All of these parallel hypersurfaces are Lie equivalent to each other and to the Legendre submanifolds induced by the Veronese surfaces.   □

CASE 2. $\rho = 0$ (*the reducible case*).

We now consider the case where ρ is identically zero. It turns out that all such Dupin submanifolds are reducible to cyclides of Dupin in $\mathbf{R}^3$. We return to the frame that we used prior to the assumption that $\rho \neq 0$. Thus, only those relations through (5.176) are valid. Since ρ is identically zero, so are all of its half-invariant derivatives. From equations (5.140) and (5.175), we see that the functions defined in equations (5.126)–(5.128) satisfy the equations

$$q = t = b = 0, \qquad a = p = 0, \qquad r = s, \qquad c = -u.$$

Thus, from equation (5.176) we have $\omega_2^7 = 0$. From these and the other relations among the Maurer–Cartan forms which we have derived, we see that the differentials of the frame vectors can be written as

$$
\begin{aligned}
dY_1 - \omega_1^1 Y_1 &= \omega_1^4 Y_4 + \omega_1^5 Y_5, \\
dY_7 - \omega_1^1 Y_7 &= \omega_7^3 Y_3 - \omega_1^5 Y_5, \\
dY_2 + \omega_1^1 Y_2 &= s(-\omega_1^4 Y_4 + \omega_1^5 Y_5), \\
dY_6 + \omega_1^1 Y_6 &= u(\omega_7^3 Y_3 + \omega_1^5 Y_5), \\
dY_3 &= \omega_7^3(-Y_6 + u Y_7), \\
dY_4 &= \omega_1^4(s Y_1 - Y_2), \\
dY_5 &= \omega_1^5(-s Y_1 - Y_2 + Y_6 - u Y_7).
\end{aligned}
\tag{5.225}
$$

Note also that from equations (5.163) and (5.164), we have

$$\tilde{d}s = ds + 2s\omega_1^1 = 0, \qquad \tilde{d}u = du + 2u\omega_1^1 = 0. \tag{5.226}$$

From equation (5.142), we have $d\omega_1^1 = 0$ and thus $d\pi = 0$. Again, we follow the procedure given after equation (5.106) to find a renormalization in which $\pi^* = 0$. Specifically, we assume that we are working in a contractible local neighborhood U, so that $\pi = d\sigma$ for some smooth scalar function σ on U. We then take a renormalization of the form (5.94) with $\tau = e^\sigma$. This results in

$$\pi^* = -\omega_1^{1*} = 0,$$

and so equation (5.225) can be rewritten as

$$
\begin{aligned}
dY_1^* &= \omega_1^{4*} Y_4 + \omega_1^{5*} Y_5, \\
dY_7^* &= \omega_7^{3*} Y_3 - \omega_1^{5*} Y_5, \\
dY_2^* &= s^*(-\omega_1^{4*} Y_4 + \omega_1^{5*} Y_5) \\
dY_6^* &= u^*(\omega_7^{3*} Y_3 + \omega_1^{5*} Y_5), \\
dY_3 &= \omega_7^{3*} Z_3^*, \\
dY_4 &= \omega_1^{4*} Z_4^*, \\
dY_5 &= \omega_1^{5*} Z_5^*,
\end{aligned}
\tag{5.227}
$$

where

$$Z_3^* = -Y_6^* - u^* Y_7^*,$$
$$Z_4^* = s^* Y_1^* - Y_2^*,$$
$$Z_5^* = -s^* Y_1^* - Y_2^* + Y_6^* - u^* Y_7^*,$$
(5.228)

and

$$s^* = \tau^{-2} s, \qquad u^* = \tau^{-2} u.$$
(5.229)

After this renormalization, half-invariant differentiation is the same as ordinary exterior differentiation. Thus, from equations (5.226) and (5.229), we have

$$ds^* = 0, \qquad du^* = 0.$$
(5.230)

i.e., s^* and u^* are constant functions on the local neighborhood U.

The frame in equation (5.227) is our final frame, and we drop the asterisks once more. Since the functions s and u are now constant, we can compute from equation (5.227) that

$$dZ_3 = -2u\omega_7^3 Y_3,$$
$$dZ_4 = 2s\omega_1^4 Y_4,$$
$$dZ_5 = 2(u - s)\omega_1^5 Y_5.$$
(5.231)

From this we see that the following 4-dimensional subspaces of $\mathbf{P}^6$,

$$\mathrm{Span}\{Y_1, Y_4, Y_5, Z_4, Z_5\},$$
$$\mathrm{Span}\{Y_7, Y_3, Y_5, Z_3, Z_5\},$$
$$\mathrm{Span}\{Y_1 + Y_7, Y_3, Y_4, Z_3, Z_4\},$$
(5.232)

are invariant under exterior differentiation, and hence they are constant. Thus, each of the three curvature sphere maps, Y_1, Y_7 and $Y_1 + Y_7$ is contained in a 4-dimensional subspace of $\mathbf{P}^6$. One can easily show that each of the subspaces in equation (5.232) has signature $(4, 1)$. Thus by Theorem 5.11, our Dupin submanifold λ on U is Lie equivalent to an open subset of a tube over a cyclide of Dupin in $\mathbf{R}^3$ in three different ways. Hence, we have the following result due to Pinkall [149].

Theorem 5.25. *Every connected Dupin submanifold $\lambda : M^3 \to \Lambda^7$ with $\rho = 0$ is reducible. It is locally Lie equivalent to a tube over a cyclide of Dupin in $\mathbf{R}^3 \subset \mathbf{R}^4$.*

Pinkall [149, p. 111] proceeds to classify Dupin submanifolds with $\rho = 0$ up to Lie equivalence. We will not do that here. The reader can follow Pinkall's proof using the fact that his constants α and β are our constants s and $-u$, respectively.

We now turn to some generalizations of this approach to higher-dimensional Dupin submanifolds. The first case that we will discuss is the case of an arbitrary proper Dupin submanifold with three curvature spheres. This case was studied in the paper of Cecil and Jensen [44]. We will briefly describe the approach of that paper here, and the reader is referred to the paper itself for the details.

We consider a connected proper Dupin submanifold $\lambda : M^{n-1} \to \Lambda^{2n-1}$ with three curvature spheres at each point. Let $\{Y_a\}$ be a smooth Lie frame on a connected open subset U of M^{n-1}, as in the case above. We can choose the Lie frame so that the three curvature spheres are Y_1, Y_{n+3} and $Y_1 + Y_{n+3}$ with respective multiplicities m_1, m_2 and m_3. Corresponding to the one function ρ in the case above, there are $m_1 m_2 m_3$ functions F_{pa}^α, where

$$1 \leq a \leq m_1,$$
$$m_1 + 1 \leq p \leq m_1 + m_2, \tag{5.233}$$
$$m_1 + m_2 + 1 \leq \alpha \leq n - 1 = m_1 + m_2 + m_3.$$

Corresponding to the case $\rho = 0$ above, we show that if there exists a fixed index, say a, such that

$$F_{pa}^\alpha = 0 \quad \text{for all } p, \alpha, \tag{5.234}$$

then the restriction of λ to the open set U is reducible. Thus, by Proposition 5.15, λ is reducible on all of M^{n-1}. Next we show that if the multiplicities are not all equal, then there exists some index a such that equation (5.234) holds, and thus λ is reducible.

Finally, we consider the case where all the multiplicities have the same value m. Using Theorem 4.19 of Chapter 4, p. 77, we show that if λ is irreducible, then λ must be Lie equivalent to the Legendre submanifold induced by an open subset of an isoparametric hypersurface in S^n. As a consequence of Cartan's [17] classification of isoparametric hypersurfaces with three principal curvatures, this implies that $m = 1, 2, 4$ or 8, and the isoparametric hypersurface must be a tube of constant radius over a standard embedding of a projective plane $\mathbf{F}P^2$ into S^{3m+1}, where $\mathbf{F}$ is the division algebra $\mathbf{R}$, $\mathbf{C}$, $\mathbf{H}$ (quaternions), $\mathbf{O}$ (Cayley numbers) for $m = 1, 2, 4, 8$, respectively.

Next we discuss the case of a proper Dupin submanifold with four curvature spheres, which was studied in the papers of Cecil and Jensen [45] and Cecil, Chi and Jensen [41]. Let $\lambda : M^{n-1} \to \Lambda^{2n-1}$ be a a connected proper Dupin submanifold with four curvature spheres at each point. Then we can find a Lie frame in which the four curvature spheres are Y_1, Y_{n+3}, $Y_1 + Y_{n+3}$ and $Y_1 + \Psi Y_{n+3}$, where Ψ is the Lie curvature of λ, having respective multiplicities m_1, m_2, m_3 and m_4. Corresponding to the one function ρ in the case above, there are four sets of functions,

$$F_{pa}^\alpha, F_{pa}^\mu, F_{\alpha a}^\mu, F_{\alpha p}^\mu, \tag{5.235}$$

where

$$1 \leq a \leq m_1,$$
$$m_1 + 1 \leq p \leq m_1 + m_2, \tag{5.236}$$
$$m_1 + m_2 + 1 \leq \alpha \leq m_1 + m_2 + m_3,$$
$$m_1 + m_2 + m_3 + 1 \leq \mu \leq n - 1 = m_1 + m_2 + m_3 + m_4.$$

As noted in Section 4.5, Thorbergsson [190] has shown for a compact proper Dupin hypersurface in S^n with four principal curvatures, the multiplicities of the principal curvatures must satisfy $m_1 = m_2, m_3 = m_4$, when the principal curvatures are appropriately ordered (see also Stolz [177]). Thus, in the papers [45] and [41], we make that assumption on the multiplicities. We also assume in [41] that the Lie curvature $\Psi = -1$, since that is true for an isoparametric hypersurface with four principal curvatures (when the principal curvatures are ordered in this way).

In [45, pp. 3–4], Cecil and Jensen conjectured that an irreducible connected proper Dupin hypersurface in S^n with four principal curvatures having multiplicities satisfying $m_1 = m_2, m_3 = m_4$ and constant Lie curvature Ψ must be Lie equivalent to an open subset of an isoparametric hypersurface in S^n.

In that paper [45], we proved that the conjecture is true if all the multiplicities are equal to one (see also Niebergall [127] who obtained the same conclusion under additional assumptions). In the second paper [41] mentioned above, we proved that the conjecture is true if $m_1 = m_2 \geq 1$, and $m_3 = m_4 = 1$, and the Lie curvature is assumed to satisfy $\Psi = -1$. We believe that the conjecture is true in its full generality, but we have not been able to prove that yet.

An important step in both papers is proving that under the assumptions on the multiplicities mentioned in the previous paragraph and with constant Lie curvature $\Psi = -1$, the Dupin submanifold λ is reducible if there exists some fixed index, say a, such that

$$F_{pa}^\alpha = F_{pa}^\mu = F_{\alpha a}^\mu = 0, \quad \text{for all } p, \alpha, \mu. \tag{5.237}$$

Thus, if λ is irreducible, no such index a can exist, and we show after a lengthy argument that λ is Lie equivalent to the Legendre submanifold induced by an open subset of an isoparametric hypersurface in S^n by invoking Theorem 4.19 of Chapter 4, p. 77.

References

[1] U. Abresch, Isoparametric hypersurfaces with four or six distinct principal curvatures, *Math. Ann.*, **264** (1983), 283–302.

[2] J. C. Álvarez Paiva, *Tautness Is Invariant Under Lie Sphere Transformations*, preprint, 2001; available online from http://www.math.poly.edu/~alvarez/pdfs/invariance.pdf.

[3] V. I. Arnold, *Mathematical Methods of Classical Mechanics*, Springer-Verlag, Berlin, 1978.

[4] E. Artin, *Geometric Algebra*, Wiley–Interscience, New York, 1957.

[5] M. F. Atiyah, R. Bott, and A. Shapiro, Clifford modules, *Topology*, **3** (1964), 3–38.

[6] T. Banchoff, The spherical two-piece property and tight surfaces in spheres, *J. Differential Geom.*, **4** (1970), 193–205.

[7] T. Banchoff and W. Kühnel, Tight submanifolds, smooth and polyhedral, in T. Cecil and S.-S. Chern, eds., *Tight and Taut Submanifolds*, MSRI Publications, Vol. 32, Cambridge University Press, Cambridge, UK, 1997, 51–118.

[8] J. Berndt, S. Console, and C. Olmos, *Submanifolds and Holonomy*, Chapman and Hall/CRC Research Notes in Mathematics, Vol. 434, Chapman and Hall, Boca Raton, FL, 2003.

[9] D. Blair, *Contact Manifolds in Riemannian Geometry*, Lecture Notes in Mathematics, Vol. 509, Springer-Verlag, Berlin, 1976.

[10] W. Blaschke, *Vorlesungen über Differentialgeometrie und geometrische Grundlagen von Einsteins Relativitätstheorie*, Vol. 3, Springer-Verlag, Berlin, 1929.

[11] G. Bol, *Projektive Differentialgeometrie*, Vol. 3, Vandenhoeck and Ruprecht, Göttingen, Netherlands, 1967.

[12] R. Bott and H. Samelson, Applications of the theory of Morse to symmetric spaces, *Amer. J. Math.*, **80** (1958), 964–1029; corrections, *Amer. J. Math.*, **83** (1961), 207–208.

[13] S. Buyske, Lie sphere transformations and the focal sets of certain taut immersions, *Trans. Amer. Math. Soc.*, **311** (1989), 117–133.

[14] M. Cahen and Y. Kerbrat, Domaines symétriques des quadriques projectives, *J. Math. Pures Appl.*, **62** (1983), 327–348.

[15] E. Cartan, *Théorie des groupes finis et continus et la géométrie différentielle traitées par la méthode du repère mobile*, Gauthiers–Villars, Paris, 1937.

[16] E. Cartan, Familles de surfaces isoparamétriques dans les espaces à courbure constante, *Ann. Mat.*, **17** (1938), 177–191.

[17] E. Cartan, Sur des familles remarquables d'hypersurfaces isoparamétriques dans les espaces sphériques, *Math. Z.*, **45** (1939), 335–367.

[18] E. Cartan, Sur quelque familles remarquables d'hypersurfaces, *C. R. Congrès Math. Liège*, (1939), 30–41.

[19] E. Cartan, Sur des familles d'hypersurfaces isoparamétriques des espaces sphériques à 5 et à 9 dimensions, *Rev. Univ. Tucuman Ser.* A, **1** (1940), 5–22.

[20] E. Cartan, *The Theory of Spinors*, Hermann, Paris, 1966; reprinted by Dover, New York, 1981.

[21] S. Carter and A. West, Tight and taut immersions, *Proc. London Math. Soc.*, **25** (1972), 701–720.

[22] S. Carter and A. West, Totally focal embeddings, *J. Differential Geom.*, **13** (1978), 251–261.

[23] S. Carter and A. West, Totally focal embeddings: Special cases, *J. Differential Geom.*, **16** (1981), 685–697.

[24] S. Carter and A. West, A characterisation of isoparametric hypersurfaces in spheres, *J. London Math. Soc.*, **26** (1982), 183–192.

[25] S. Carter and A. West, Convexity and cylindrical two-piece properties, *Illinois J. Math.*, **29** (1985), 39–50.

[26] S. Carter and A. West, Isoparametric systems and transnormality, *Proc. London Math. Soc.*, **51** (1985), 520–542.

[27] S. Carter and A. West, Generalized Cartan polynomials, *J. London Math. Soc.*, **32** (1985), 305–316.

[28] S. Carter and A. West, Isoparametric and totally focal submanifolds, *Proc. London Math. Soc.*, **60** (1990), 609–624.

[29] A. Cayley, On the cyclide, *Quart. J. Pure Appl. Math.*, **12** (1873), 148–165; reprinted in *Collected Mathematical Papers*, Vol. 9, Cambridge University Press, Cambridge, UK, 1896, 64–78.

[30] T. Cecil, A characterization of metric spheres in hyperbolic space by Morse theory, *Tôhoku Math. J.*, **26** (1974), 341–351.

[31] T. Cecil, Taut immersions of non-compact surfaces into a Euclidean 3-space, *J. Differential Geom.*, **11** (1976), 451–459.

[32] T. Cecil, Reducible Dupin submanifolds, *Geom. Dedicata*, **32** (1989), 281–300.

[33] T. Cecil, On the Lie curvatures of Dupin hypersurfaces, *Kodai Math. J.*, **13** (1990), 143–153.

[34] T. Cecil, Lie sphere geometry and Dupin submanifolds, in *Geometry and Topology of Submanifolds* III, World Scientific, River Edge, NJ, 1991, 90–107.

[35] T. Cecil, Dupin submanifolds, in *Geometry and Topology of Submanifolds* V, World Scientific, River Edge, NJ, 1993, 77–102.

[36] T. Cecil, Taut and Dupin submanifolds, in T. Cecil and S.-S. Chern, eds., *Tight and Taut Submanifolds*, MSRI Publications, Vol. 32, Cambridge University Press, Cambridge, UK, 1997, 135–180.

[37] T. Cecil and S.-S. Chern, Tautness and Lie sphere geometry, *Math. Ann.*, **278** (1987), 381–399.

[38] T. Cecil and S.-S. Chern, Dupin submanifolds in Lie sphere geometry, in *Differential Geometry and Topology: Proceedings, Tianjin*, 1986–87, Lecture Notes in Mathematics, Vol. 1369, Springer-Verlag, Berlin, New York, 1989, 1–48.

[39] T. Cecil and S.-S. Chern, eds., *Tight and Taut Submanifolds*, MSRI Publications, Vol. 32, Cambridge University Press, Cambridge, UK, 1997.

[40] T. Cecil, Q.-S. Chi, and G. Jensen, Isoparametric hypersurfaces with four principal curvatures, *Ann. Math.*, **166** (2007), 1–76.

[41] T. Cecil, Q.-S. Chi, and G. Jensen, Dupin hypersurfaces with four principal curvatures II, *Geom. Dedicata*, to appear.

[42] T. Cecil, Q.-S. Chi, and G. Jensen, *On Kuiper's Conjecture*, preprint, 2007, math/0512089.

[43] T. Cecil and G. Jensen, Dupin hypersurfaces, *Geometry and Topology of Submanifolds* VII, World Scientific, River Edge, NJ, 1995, 100–105.

[44] T. Cecil and G. Jensen, Dupin hypersurfaces with three principal curvatures, *Invent. Math.*, **132** (1998), 121–178.

[45] T. Cecil and G. Jensen, Dupin hypersurfaces with four principal curvatures, *Geom. Dedicata*, **79** (2000), 1–49.

[46] T. Cecil and P. Ryan, Focal sets of submanifolds, *Pacific J. Math.*, **78** (1978), 27–39.

[47] T. Cecil and P. Ryan, Focal sets, taut embeddings and the cyclides of Dupin, *Math. Ann.*, **236** (1978), 177–190.

[48] T. Cecil and P. Ryan, Distance functions and umbilical submanifolds of hyperbolic space, *Nagoya Math. J.*, **74** (1979), 67–75.

[49] T. Cecil and P. Ryan, Tight and taut immersions into hyperbolic space, *J. London Math. Soc.*, **19** (1979), 561–572.

[50] T. Cecil and P. Ryan, Conformal geometry and the cyclides of Dupin, *Canadian J. Math.*, **32** (1980), 767–782.

[51] T. Cecil and P. Ryan, *Tight Spherical Embeddings*, Lecture Notes in Mathematics, Vol. 838, Springer-Verlag, Berlin, New York, 1981, 94–104.

[52] T. Cecil and P. Ryan, *Tight and Taut Immersions of Manifolds*, Research Notes in Mathematics, Vol. 107, Pitman, London, 1985.

[53] S.-S. Chern, An introduction to Dupin submanifolds, in *Differential Geometry*, Pitman Monographs Surveys in Pure and Applied Mathematics, Vol. 52, Longman Scientific and Technical, Harlow, UK, 1991, 95–102.

[54] L. Conlon, *Differentiable Manifolds: A First Course*, 1st ed., Birkhäuser Boston, Cambridge, MA, 1993.

[55] M. Dajczer, L. Florit, and R. Tojeiro, Reducibility of Dupin submanifolds, *Illinois J. Math.*, **49** (2005), 759–791.

[56] G. Darboux, *Leçons sur la théorie générale des surfaces*, 2nd ed., Gauthiers–Villars, Paris, 1941.

[57] W. Degen, Generalized cyclides for use in CAGD, in *Computer-Aided Surface Geometry and Design (Bath, 1990)*, Institute of Mathematics and Its Applications Conference Series, New Series, Vol. 48, Oxford University Press, New York, 1994, 349–363.

[58] J. Dorfmeister and E. Neher, An algebraic approach to isoparametric hypersurfaces I, II, *Tôhoku Math. J.*, **35** (1983), 187–224, 225–247.

[59] J. Dorfmeister and E. Neher, Isoparametric triple systems of algebra type, *Osaka J. Math.*, **20** (1983), 145–175.

[60] J. Dorfmeister and E. Neher, Isoparametric triple systems of FKM-type I, *Abh. Math. Sem. Hamburg*, **53** (1983), 191–216.

[61] J. Dorfmeister and E. Neher, Isoparametric triple systems of FKM-type II, *Manuscripta Math.*, **43** (1983), 13–44.

[62] J. Dorfmeister and E. Neher, Isoparametric hypersurfaces, case $g = 6, m = 1$, *Comm. Algebra*, **13** (1985), 2299–2368.

[63] J. Dorfmeister and E. Neher, Isoparametric triple systems with special Z-structure, *Algebras Groups Geom.*, **7** (1990), 21–94.

[64] C. Dupin, *Applications de géométrie et de méchanique: À la marine, aux points et chaussées, etc.*, Bachelier, Paris, 1822.

[65] S. Eilenberg and N. Steenrod, *Foundations of Algebraic Topology*, Princeton University Press, Princeton, NJ, 1952.

[66] L. Eisenhart, *A Treatise on the Differential Geometry of Curves and Surfaces*, Ginn, Boston, 1909.

[67] F. Fang, *Multiplicities of Principal Curvatures of Isoparametric Hypersurfaces*, preprint, Max Planck Institut für Mathematik, Bonn, Germany, 1996.

[68] F. Fang, On the topology of isoparametric hypersurfaces with four distinct principal curvatures, *Proc. Amer. Math. Soc.*, **127** (1999), 259–264.

[69] F. Fang, Topology of Dupin hypersurfaces with six principal curvatures, *Math. Z.*, **231** (1999), 533–555.

[70] E. V. Ferapontov, Dupin hypersurfaces and integrable hamiltonian systems of hydrodynamic type, which do not possess Riemann invariants, *Differential Geom. Appl.*, **5** (1995), 121–152.

[71] E. V. Ferapontov, Isoparametric hypersurfaces in spheres, integrable nondiagonalizable systems of hydrodynamic type, and N-wave systems, *Differential Geom. Appl.*, **5** (1995), 335–369.

[72] E. V. Ferapontov, Lie sphere geometry and integrable systems, *Tôhoku Math. J.*, **52** (2000), 199–233.

[73] D. Ferus, H. Karcher, and H.-F. Münzner, Cliffordalgebren und neue isoparametrische Hyperflächen, *Math. Z.*, **177** (1981), 479–502.

[74] J. Fillmore, The fifteen-parameter conformal group, *Internat. J. Theoret. Phys.*, **16** (1977), 937–963.

[75] J. Fillmore, On Lie's higher sphere geometry, *Enseign. Math.*, **25** (1979), 77–114.

[76] J. Fillmore, Hermitean quadrics as contact manifolds, *Rocky Mountain J. Math.*, **14** (1984), 559–571.

[77] K. Fladt and A. Baur, *Analytische Geometrie spezieller Flächen und Raumkurven*, Vieweg, Braunschweig, Germany, 1975.

[78] P. Griffiths, On Cartan's method of Lie groups and moving frames as applied to uniqueness and existence questions in differential geometry, *Duke Math. J.*, **41** (1974), 775–814.

[79] K. Grove and S. Halperin, Dupin hypersurfaces, group actions, and the double mapping cylinder, *J. Differential Geom.*, **26** (1987), 429–459.

[80] J. Hahn, Isoparametric hypersurfaces in the pseudo-Riemannian space forms, *Math. Z.*, **187** (1984), 195–208.

[81] J. Hahn, Isotropy representations of semisimple symmetric spaces and homogeneous hypersurfaces, *J. Math. Soc. Japan*, **40** (1988), 271–288.

[82] C. Harle, Isoparametric families of submanifolds, *Bol. Soc. Brasil Mat.*, **13** (1982), 35–48.

[83] T. Hawkins, Line geometry, differential equations and the birth of Lie's theory of groups, in *The History of Modern Mathematics*, Vol. I, Academic Press, San Diego, 1989, 275–327.

[84] J. Hebda, Manifolds admitting taut hyperspheres, *Pacific J. Math.*, **97** (1981), 119–124.

[85] J. Hebda, Some new tight embeddings which cannot be made taut, *Geom. Dedicata*, **17** (1984), 49–60.

[86] J. Hebda, The possible cohomology of certain types of taut submanifolds, *Nagoya Math. J.*, **111** (1988), 85–97.

[87] E. Heintze and X. Liu, Homogeneity of infinite dimensional isoparametric submanifolds, *Ann. Math.*, **149** (1999), 149–181.

[88] E. Heintze, C. Olmos, and G. Thorbergsson, Submanifolds with constant principal curvatures and normal holonomy groups, *Internat. J. Math.*, **2** (1991), 167–175.

[89] U. Hertrich-Jeromin, *Introduction to Möbius Differential Geometry*, London Mathematical Society Lecture Notes Series, Vol. 300, Cambridge University Press, Cambridge, UK, 2003.

[90] D. Hilbert and S. Cohn-Vossen, *Geometry and the Imagination*, Chelsea, New York, 1952.

[91] W.-Y. Hsiang, R. Palais, and C.-L. Terng, The topology of isoparametric submanifolds, *J. Differential Geom.*, **27** (1988), 423–460.

[92] S. Immervoll, On the classification of isoparametric hypersurfaces with four distinct principal curvatures in spheres, *Ann. Math.*, to appear.

[93] G. Jensen, *Higher Order Contact of Submanifolds of Homogeneous Spaces*, Lecture Notes in Mathematics, Vol. 610, Springer-Verlag, Berlin, 1977.

[94] F. Klein, *Vorlesungen über höhere Geometrie*, Springer-Verlag, Berlin, 1926; reprinted by Chelsea, New York, 1957.

[95] S. Kobayashi and K. Nomizu, *Foundations of Differential Geometry*, Vols. I and II, Wiley–Interscience, New York, 1963, 1969.

[96] N. Kuiper, On convex maps, *Nieuw Arch. Wisk.*, **10** (1962), 147–164.

[97] N. Kuiper, Minimal total absolute curvature for immersions, *Invent. Math.*, **10** (1970), 209–238.

[98] N. Kuiper, Tight embeddings and maps: Submanifolds of geometrical class three in E^n, in *The Chern Symposium 1979* (*Proceedings of the International Symposium, Berkeley, CA, 1979*), Springer-Verlag, Berlin, New York, 1980, 97–145.

[99] N. Kuiper, Taut sets in three space are very special, *Topology*, **23** (1984), 323–336.

[100] E. P. Lane, *A Treatise on Projective Differential Geometry*, University of Chicago Press, Chicago, 1942.

[101] T. Levi-Civita, Famiglie di superficie isoparametrische nell'ordinario spacio euclideo, *Atti. Accad. Naz. Lincei. Rend. Cl. Sci. Fis. Mat. Natur.*, **26** (1937), 355–362.

[102] T. Z. Li, Laguerre geometry of hypersurfaces in $\mathbf{R}^n$, *Manuscripta Math.*, **122** (2007), 73–95.

[103] T. Z. Li and C.-P. Wang, Laguerre geometry of surfaces in $\mathbf{R}^3$, *Acta Math. Sinica*, **21** (2005), 1525–1534.

[104] S. Lie, Über Komplexe, inbesondere Linien- und Kugelkomplexe, mit Anwendung auf der Theorie der partieller Differentialgleichungen, *Math. Ann.*, **5** (1872), 145–208, 209–256 (Ges. Abh. 2, 1–121).

[105] S. Lie and G. Scheffers, *Geometrie der Berührungstransformationen*, Teubner, Leipzig, 1896; reprinted by Chelsea, New York, 1977.

[106] R. Lilienthal, Besondere Flächen, in *Encyklopädie der Mathematische Wissenschaften*, Vol. III, Teubner, Leipzig, 1902–1927, 269–354.

[107] J. Liouville, Note au sujet de l'article précedént, *J. Math. Pure Appl.* (1), **12** (1847), 265–290.

[108] M. Magid, Lorentzian isoparametric hypersurfaces, *Pacific J. Math.*, **118** (1985), 165–197.

[109] J. C. Maxwell, On the cyclide, *Quart. J. Pure Appl. Math.*, **34** (1867); reprinted in *Collected Works*, Vol. 2, Dover, New York, 1952, 144–159.

[110] J. Milnor, *Morse Theory*, Annals of Mathematics Studies, Vol. 51, Princeton University Press, Princeton, NJ, 1963.

[111] R. Miyaoka, Compact Dupin hypersurfaces with three principal curvatures, *Math. Z.*, **187** (1984), 433–452.

[112] R. Miyaoka, Taut embeddings and Dupin hypersurfaces, in *Differential Geometry of Submanifolds: Proceedings of the Conference in Kyoto, Japan, 1984*, Lecture Notes in Mathematics, Vol. 1090, Springer-Verlag, Berlin, New York, 1984, 15–23.

[113] R. Miyaoka, Dupin hypersurfaces and a Lie invariant, *Kodai Math. J.*, **12** (1989), 228–256.

[114] R. Miyaoka, Dupin hypersurfaces with six principal curvatures, *Kodai Math. J.*, **12** (1989), 308–315.

[115] R. Miyaoka, Lie contact structures and normal Cartan connections, *Kodai Math. J.*, **14** (1991), 13–41.

[116] R. Miyaoka, Lie contact structures and conformal structures, *Kodai Math. J.*, **14** (1991), 42–71.

[117] R. Miyaoka, A note on Lie contact manifolds, in *Progress in Differential Geometry*, Advanced Studies in Pure Mathematics, Vol. 22, Mathematical Society of Japan, Tokyo, 1993, 169–187.

[118] R. Miyaoka, The linear isotropy group of $G_2/SO(4)$, the Hopf fibering and isoparametric hypersurfaces, *Osaka J. Math.*, **30** (1993), 179–202.

[119] R. Miyaoka, A new proof of the homogeneity of isoparametric hypersurfaces with $(g, m) = (6, 1)$, in *Geometry and Topology of Submanifolds* X, World Scientific, River Edge, NJ, 2000, 178–199.

[120] R. Miyaoka, *The Dorfmeister–Neher Theorem on Isoparametric Hypersurfaces*, preprint, 2006, math.DG/0602519.

[121] R. Miyaoka and T. Ozawa, Construction of taut embeddings and Cecil–Ryan conjecture, in K. Shiohama, ed., *Geometry of Manifolds*, Perspectives in Mathematics, Vol. 8, Academic Press, New York, 1989, 181–189.

[122] M. Morse and S. Cairns, *Critical Point Theory in Global Analysis and Differential Topology*, Academic Press, New York, 1969.

[123] H.-F. Münzner, Isoparametrische Hyperflächen in Sphären I, II, *Math. Ann.*, **251** (1980), 57–71, **256** (1981), 215–232.

[124] E. Musso and L. Nicolodi, A variational problem for surfaces in Laguerre geometry, *Trans. Amer. Math. Soc.*, **348** (1996), 4321–4337.

[125] E. Musso and L. Nicolodi, Laguerre geometry for surfaces with plane lines of curvature, *Abh. Math. Sem. Univ. Hamburg*, **69** (1999), 123–138.

[126] R. Niebergall, Dupin hypersurfaces in $\mathbf{R}^5$ I, *Geom. Dedicata*, **40** (1991), 1–22.

[127] R. Niebergall, Dupin hypersurfaces in $\mathbf{R}^5$ II, *Geom. Dedicata*, **41** (1992), 5–38.

[128] R. Niebergall and P. Ryan, Isoparametric hypersurfaces: The affine case, in *Geometry and Topology of Submanifolds* V, World Scientific, River Edge, NJ, 1993, 201–214.

[129] R. Niebergall and P. Ryan, Affine isoparametric hypersurfaces, *Math. Z.*, **217** (1994), 479–485.

[130] R. Niebergall and P. Ryan, Focal sets in affine geometry, in *Geometry and Topology of Submanifolds* VI, World Scientific, River Edge, NJ, 1994, 155–164.

[131] R. Niebergall and P. Ryan, Affine Dupin surfaces, *Trans. Amer. Math. Soc.*, **348** (1996), 1093–1117.

[132] R. Niebergall and P. Ryan, Real hypersurfaces in complex space forms, in T. Cecil and S.-S. Chern, eds., *Tight and Taut Submanifolds*, MSRI Publications, Vol. 32, Cambridge University Press, Cambridge, UK, 1997, 233–305.

[133] K. Nomizu, *Fundamentals of Linear Algebra*, McGraw–Hill, New York, 1966.

[134] K. Nomizu, Characteristic roots and vectors of a differentiable family of symmetric matrices, *Linear Multilinear Algebra*, **2** (1973), 159–162.

[135] K. Nomizu, Some results in E. Cartan's theory of isoparametric families of hypersurfaces, *Bull. Amer. Math. Soc.*, **79** (1973), 1184–1188.

[136] K. Nomizu, Élie Cartan's work on isoparametric families of hypersurfaces, in S. S. Chern and R. Osserman, eds., *Differential Geometry*, Part 1, Proceedings of Symposia in Pure Mathematics, Vol. 27, American Mathematical Society, Providence, RI, 1975, 191–200.

[137] K. Nomizu, On isoparametric hypersurfaces in Lorentzian space forms, *Japan. J. Math.*, **7** (1981), 217–226.

[138] K. Nomizu and L. Rodriguez, Umbilical submanifolds and Morse functions, *Nagoya Math. J.*, **48** (1972), 197–201.

[139] C. Olmos, Isoparametric submanifolds and their homogeneous structure, *J. Differential Geom.*, **38** (1993), 225–234.

[140] C. Olmos, Homogeneous submanifolds of higher rank and parallel mean curvature, *J. Differential Geom.*, **39** (1994), 605–627.

[141] B. O'Neill, *Semi-Riemannian Geometry*, Academic Press, New York, 1983.

[142] T. Ozawa, On the critical sets of distance functions to a taut submanifold, *Math. Ann.*, **276** (1986), 91–96.

[143] H. Ozeki and M. Takeuchi, On some types of isoparametric hypersurfaces in spheres I, II, *Tôhoku Math. J.*, **27** (1975), 515–559, **28** (1976), 7–55.

[144] R. Palais and C.-L. Terng, A general theory of canonical forms, *Trans. Amer. Math. Soc.*, **300** (1987), 771–789.

[145] R. Palais and C.-L. Terng, *Critical Point Theory and Submanifold Geometry*, Lecture Notes in Mathematics, Vol. 1353, Springer-Verlag, Berlin, New York, 1988.

[146] U. Pinkall, *Dupin'sche Hyperflächen*, Dissertation, Universität Freiburg, Freiburg, Germany, 1981.

[147] U. Pinkall, *W*-Kurven in der ebenen Lie-Geometrie, *Elemente Math.*, **39** (1984), 28–33, 67–78.

[148] U. Pinkall, Letter to T. Cecil, December 5, 1984.

[149] U. Pinkall, Dupin'sche Hyperflächen in E^4, *Manuscripta Math.*, **51** (1985), 89–119.

[150] U. Pinkall, Dupin hypersurfaces, *Math. Ann.*, **270** (1985), 427–440.

[151] U. Pinkall, Curvature properties of taut submanifolds, *Geom. Dedicata*, **20** (1986), 79–83.

[152] U. Pinkall and G. Thorbergsson, Deformations of Dupin hypersurfaces, *Proc. Amer. Math. Soc.*, **107** (1989), 1037–1043.

[153] U. Pinkall and G. Thorbergsson, Taut 3-manifolds, *Topology*, **28** (1989), 389–401.

[154] U. Pinkall and G. Thorbergsson, Examples of infinite dimensional isoparametric submanifolds, *Math. Z.*, **205** (1990), 279–286.

[155] M. J. Pratt, Cyclides in computer aided geometric design, *Comput. Aided Geom. Design*, **7** (1990), 221–242.

[156] M. J. Pratt, Cyclides in computer aided geometric design II, *Comput. Aided Geom. Design*, **12** (1995), 131–152.

[157] H. Reckziegel, Krümmungsflächen von isometrischen Immersionen in Räume konstanter Krümmung, *Math. Ann.*, **223** (1976), 169–181.

[158] H. Reckziegel, On the eigenvalues of the shape operator of an isometric immersion into a space of constant curvature, *Math. Ann.*, **243** (1979), 71–82.

[159] H. Reckziegel, Completeness of curvature surfaces of an isometric immersion, *J. Differential Geom.*, **14** (1979), 7–20.

[160] C. M. Riveros and K. Tenenblat, On four dimensional Dupin hypersurfaces in Euclidean space, *An. Acad. Brasil Ciênc.*, **75** (2003), 1–7.

[161] C. M. Riveros and K. Tenenblat, Dupin hypersurfaces in $\mathbf{R}^5$, *Canadian J. Math.*, **57** (2005), 1291–1313.

[162] D. Rowe, The early geometrical works of Sophus Lie and Felix Klein, in *The History of Modern Mathematics*, Vol. I, Academic Press, San Diego, 1989, 209–273.

[163] P. Ryan, Homogeneity and some curvature conditions for hypersurfaces, *Tôhoku Math. J.*, **21** (1969), 363–388.

[164] P. Ryan, Hypersurfaces with parallel Ricci tensor, *Osaka J. Math.*, **8** (1971), 251–259.

[165] P. Ryan, *Euclidean and Non-Euclidean Geometry*, Cambridge University Press, Cambridge, UK, 1986.

[166] P. Samuel, *Projective Geometry*, Springer-Verlag, Berlin, 1988.

[167] S. Sasaki and T. Suguri, On the problems of equivalence of plane curves in the Lie's higher circle geometry and minimal curves in the conformal geometry, *Tôhoku Math. J.*, **47** (1940), 77–86.

[168] M. Schrott and B. Odehnal, Ortho-circles of Dupin cyclides, *J. Geom. Graph.*, **10** (2006), 73–98.

[169] B. Segre, Famiglie di ipersuperficie isoparametrische negli spazi euclidei ad un qualunque numero di demesioni, *Atti. Accad. Naz. Lincei. Rend. Cl. Sci. Fis. Mat. Natur.*, **27** (1938), 203–207.

[170] D. Singley, Smoothness theorems for the principal curvatures and principal vectors of a hypersurface, *Rocky Mountain J. Math.*, **5** (1975), 135–144.

[171] M. Spivak, *A Comprehensive Introduction to Differential Geometry*, Vols. 1–5, Publish or Perish, Boston, 1970–1975.

[172] Y. L. Srinivas and D. Dutta, Blending and joining using cyclides, *ASME Trans. J. Mechanical Design*, **116** (1994), 1034–1041.

[173] Y. L. Srinivas and D. Dutta, An intuitive procedure for constructing complex objects using cyclides, *CAD*, **26** (1994), 327–335.

[174] Y. L. Srinivas and D. Dutta, Cyclides in geometric modeling: Computational tools for an algorithmic infrastructure, *ASME Trans. J. Mechanical Design*, **117** (1995), 363–373.

[175] Y. L. Srinivas and D. Dutta, Rational parametrization of parabolic cyclides, *Comput. Aided Geom. Design*, **12** (1995), 551–566.

[176] S. Sternberg, *Lectures on Differential Geometry*, Prentice–Hall, Englewood Cliffs, 1964; 2nd ed., Chelsea, New York, 1983.

[177] S. Stolz, Multiplicities of Dupin hypersurfaces, *Invent. Math.*, **138** (1999), 253–279.

[178] W. Strübing, Isoparametric submanifolds, *Geom. Dedicata*, **20** (1986), 367–387.

[179] R. Takagi, A class of hypersurfaces with constant principal curvatures in a sphere, *J. Differential Geom.*, **11** (1976), 225–233.

[180] R. Takagi and T. Takahashi, On the principal curvatures of homogeneous hypersurfaces in a sphere, in *Differential Geometry: In Honor of K. Yano*, Kinokuniya, Tokyo, 1972, 469–481.

[181] M. Takeuchi, Proper Dupin hypersurfaces generated by symmetric submanifolds, *Osaka Math. J.*, **28** (1991), 153–161.

[182] M. Takeuchi and S. Kobayashi, Minimal imbeddings of *R*-spaces, *J. Differential Geom.*, **2** (1968), 203–215.

[183] C.-L. Terng, Isoparametric submanifolds and their Coxeter groups, *J. Differential Geom.*, **21** (1985), 79–107.

[184] C.-L. Terng, Convexity theorem for isoparametric submanifolds, *Invent. Math.*, **85** (1986), 487–492.

[185] C.-L. Terng, Submanifolds with flat normal bundle, *Math. Ann.*, **277** (1987), 95–111.

[186] C.-L. Terng, Proper Fredholm submanifolds of Hilbert space, *J. Differential Geom.*, **29** (1989), 9–47.

[187] C.-L. Terng, *Recent Progress in Submanifold Geometry*, Part 1, Proceedings of Symposia in Pure Mathematics, Vol. 54, American Mathematical Society, Providence, RI, 1993, 439–484.

[188] C.-L. Terng and G. Thorbergsson, Submanifold geometry in symmetric spaces, *J. Differential Geom.*, **42** (1995), 665–718.

[189] C.-L. Terng and G. Thorbergsson, Taut immersions into complete Riemannian manifolds, in T. Cecil and S.-S. Chern, eds., *Tight and Taut Submanifolds*, MSRI Publications, Vol. 32, Cambridge University Press, Cambridge, UK, 1997, 181–228.

[190] G. Thorbergsson, Dupin hypersurfaces, *Bull. London Math. Soc.*, **15** (1983), 493–498.

[191] G. Thorbergsson, Isoparametric foliations and their buildings, *Ann. Math.*, **133** (1991), 429–446.

[192] G. Thorbergsson, A survey on isoparametric hypersurfaces and their generalizations, in *Handbook of Differential Geometry*, Vol. I, North-Holland, Amsterdam, 2000, 963–995.

[193] K. Voss, *Eine Verallgemeinerung der Dupinsche Zykliden*, Tagungsbericht 41/1981, Geometrie, Mathematisches Forschungsinstitut Oberwolfach, Oberwolfach, Germany, 1981.

[194] C.-P. Wang, Surfaces in Möbius geometry, *Nagoya Math. J.*, **125** (1992), 53–72.

[195] C.-P. Wang, Möbius geometry for hypersurfaces in S^4, *Nagoya Math. J.*, **139** (1995), 1–20.

[196] C.-P. Wang, Möbius geometry of submanifolds in S^n, *Manuscripta Math.*, **96** (1998), 517–534.

[197] Q.-M. Wang, Isoparametric functions on Riemannian manifolds I, *Math. Ann.*, **277** (1987), 639–646.

[198] Q.-M. Wang, On the topology of Clifford isoparametric hypersurfaces, *J. Differential Geom.*, **27** (1988), 55–66.

[199] A. West, Isoparametric systems, in *Geometry and Topology of Submanifolds* I, World Scientific, River Edge, NJ, 1989, 222–230.

[200] A. West, Isoparametric systems on symmetric spaces, in *Geometry and Topology of Submanifolds* V, World Scientific, River Edge, NJ, 1993, 281–287.

[201] B. Wu, Isoparametric submanifolds of hyperbolic spaces, *Trans. Amer. Math. Soc.*, **331** (1992), 609–626.

[202] B. Wu, A finiteness theorem for isoparametric hypersurfaces, *Geom. Dedicata*, **50** (1994), 247–250.

[203] Q. Zhao, Isoparametric submanifolds of hyperbolic spaces, *Chinese J. Contemp. Math.*, **14** (1993), 339–346.

Index